Writing Better
Technical
Articles

Writing Better Technical Articles

Harley Bjelland

TAB BOOKS

Blue Ridge Summit, PA

Trademarks

CompuServe	H&R Block
Delphi	
Dialog	Lockheed Missles
GEnie	General Electric
IBM®	International Business Machines Corp.®
Knowledge Index	Lockheed Missles
Mead Data Central	Mead Corp
Newsnet	
Orbit	SDC/Orbit
Videolog	Schweber Electronics
Vu/Text	
Wilsonline	H.W. Wilson
WordStar®	
WordPerfect	
XYWrite™	
Yellow Pages	Dun & Bradstreet

FIRST EDITION
FIRST PRINTING

Library of Congress Cataloging-in-Publication Data

Bjelland, Harley.
   Writing better technical articles / by Harley Bjelland.
      p.   cm.
   ISBN 0-8306-8439-5   ISBN 0-8306-3439-8 (pbk.)
   1. Technical writing.   I. Title.
  T11.B54   1990
  808'.0666—dc20   89-49448
                                                CIP

TAB BOOKS offers software for sale. For information and a catalog, please contact TAB Software Department, Blue Ridge Summit, PA 17294-0850.

Questions regarding the content of this book should be addressed to:

   Reader Inquiry Branch
   TAB BOOKS
   Blue Ridge Summit, PA 17294-0214

Acquisitions Editor: Roland S. Phelps
Technical Editor: David M. Gauthier
Production: Katherine Brown
Book Design: Jaclyn J. Boone

Contents

Acknowledgments

My wife, Tokiko (AKA Dorrie), has been infinitely patient during the long inspiration, gestation, and perspiration periods required for this book. And she served well as my "guinea pig" in making my text more useable.

Many thanks to Roland Phelps, Electronics Acquisitions Editor at TAB BOOKS who helped me along the way and who believed in the book.

Floyd Ashburn, owner of Compatibility + Plus in Springfield, Oregon, supplied me with an excellent IBM XT and with much sage advice and assistance along the way. Thanks Floyd.

I'm grateful for the excellent assistance the people at the Eugene, Oregon, library gave me in finding obscure facts that I did not even know existed.

And thanks to Thor, my computer, who took so much of the drudgery out of writing and made it a creative experience.

To the light of my life,
my wife,
Tokiko.

Introduction

This book is targeted first at showing how to write better technical articles, since this is the easiest and best way for a technical professional to become published for the first time. However, all of the techniques taught in this book apply to all types of technical writing, including books, memos, technical manuals, proposals, letters, etc. So the book can be used as a basic text applying to all forms of technical writing.

This book was designed to be read from beginning to end. The chapters form a series of activities which show you, step-by-step, how to get published. Each chapter builds and expands on the material in the previous chapter(s).

To use this book, first read the Table of Contents. This gives you a quick outline of the entire book. Next, browse through the book, look at the headlines, the section headings and the visuals to get a better "picture" of the book's contents and the manner in which information is presented.

Finally, start with Chapter 1 and read through all of the other chapters, in order. Study the examples and use the figures and formulas provided to create your own letters/memos and reports. Use the Precedent Sort to organize your writing.

Above all, be an active reader. When you read, keep a marking pen nearby and underline or highlight the points you especially want to remember. This activity reinforces the information in your mind and stores it in your memory bank for later recall. And it helps you locate the information later.

Don't wait until you have read the entire book to start practicing the writing techniques taught here. Remember the aphorism, "What I read, I forget. What I see (visuals), I remember. What I do, I understand."

As you read through the book, try writing short paragraphs, a memo, a query letter. Practice organizing topics for an article. Make a sketch of something and write a description of how it works. Don't worry, no one is going to read it, this is just practice. Visit the library and get acquainted with the marvelous references available.

This book was not written just to be *read*. This book was written to be *used*. So, use it well and you'll find that writing of things technical will indeed become easy, and rewarding in many ways.

Although the title of this book is *Writing Better Technical Articles*, all of the principles taught will also help you write better memos, letters, reports, proposals, manuals—all of the day-to-day writing you have to do in your profession.

In science, agriculture, bacteriology, chemistry, forestry, engineering, mining, medicine, metallurgy, physics, or any other technical field, writing is essential in most stages of every important project if a person does not want to remain anonymous.

Chapter 1 gives an overview of your first step, to select a good subject, one that suits both you and your audience. If you start out in the wrong direction, you'll waste a lot of valuable time and effort. Chapter 2 shows you how to get started right with a good subject.

Once you've selected a good subject, you should perform your brief, preliminary research, then write a query letter to the periodical(s) of your choice. Chapter 3 covers this important aspect of writing. This step is very important. You will save a lot of your valuable time and effort if you query a periodical before you write the article so you can become familiar with that periodical's specific editorial requirements and their need for articles. Some periodicals are totally staff-written, so it would be a waste of your time to try to publish in such periodicals.

After an editor's go-ahead, complete your in-depth research as discussed in chapter 4. Some of the information you need may be in the library, so this chapter, with appendix A, gives you a comprehensive, up-to-date refresher course in how to use a library.

The most important single key to written communication is: Organization

An article, properly organized, is half written. In chapter 5 you'll learn how to take your raw data and your visuals, and organize them using the *Precedent Sort*. The precedent sort is a unique technique, a "formula" that I have adapted, tested, and refined over many years and used in all types of publications, ranging from simple memos to full-length books.

Briefly the precedent sort shows you how to create and organize a random group of topics so that they form a logical, smooth-flowing, coherent, well-written narrative. The precedent sort helps you develop the proper and logical sequence to use to effectively communicate your ideas to others.

For the next step, add your visuals, as described in chapter 6. It's also very important to plan and sketch your visuals before you write so your illustrations form an integral part of your text, instead of being tacked on as afterthoughts. Visuals can simplify your writing since each picture can save hundreds of words. Proper visuals also make technical concepts much easier to understand than words do.

Once your visuals have been properly integrated with your outline, it's time to write your first draft. Chapter 7 gives an efficient, relatively painless, orderly method of doing so.

One of the most important uses of the modern personal computer is its application as a word processor. It has revolutionized the way we write. No longer do we have to write, rewrite, and retype time and time again to come up with a perfect manuscript. Word processing programs store all the text we input so we need to edit only those portions that need changing to arrive at a complete draft.

And most word processing programs operate with companion spelling programs, which you can expand to include the special words or jargon of your technical profession to a supplemental electronic dictionary. In addition, a thesaurus is available as an adjunct to many word processing programs to help select alternate words to better express your ideas. Some special programs check the grammatical aspects of your writing, including sentence length, percentage of long words, overusage of the passive voice, etc.

The age of computer-assisted writing is indeed here and promises to do even more in the future to take much of the drudgery out of this important task of communicating. Chapter 7 describes how to effectively use the important capabilities of modern word processors and auxiliary programs to enhance, simplify, and make your writing easier.

Chapter 8 shows you how to add a beginning and an end to your article, and how to prepare your manuscript for submittal to a periodical. Also included is a comprehensive summary of the survey I have made of the editorial requirements, both current and future, of over 75 periodicals in all branches of engineering and science.

For the last step, you mail your manuscript off to the editor. But don't sit and wait for your acceptance check to come in and for your article to appear in print. Start researching another article immediately; before you realize it, you'll have a half dozen articles ready to list on your resumé and you'll be on your way to writing a full-length book.

Chapter 9 covers the revolution that is occurring in on-line systems and shows how to research the thousands of databases located throughout the United States and in many foreign countries. This exploding field promises to radically change the manner in which we do research. Printed books may soon be on the way out since a single 256-megabyte optical storage disk can store the equivalent of 300 to 400 books. With a personal computer, a modem, and a telephone line, you can research gigantic databases from your home or office.

Introduction

There are millions of pages of information in the thousands of information repositories located throughout the world.

Chapter 10 is a bonus chapter — a post-graduate course for writers who are serious about improving their writing abilities even more. You'll learn about readability principles and techniques that make your writing easier to read and understand, and to increase the chances of your getting published.

1

Why You Should Write: Publish and Flourish

Your knowing is nothing, unless others know you know.

Persius

Forbes magazine asked a number of successful corporate executives what people should learn to help prepare them for careers. Their answer:

Teach them to write better.

The primary goals of this book are to meet the *Forbes'* recommendation and to dispel the notion that engineers and scientists cannot write. This book is going to teach technical people how to write better and with less effort.

This book will help overcome the overwhelming and largely unnecessary emphasis on style, spelling, grammar, and the Roman-numeral outlines that discouraged you in your high school and college English classes. This practical text will teach you how to write more proficiently using new effective writing formulas and techniques that have never before been applied to technical writing.

THE ANONYMOUS ENGINEER

Kevin Hansen worked diligently as a design engineer for a number of years. Few members of upper management knew his name. No one criticized or complimented him for his accomplishments. Quiet and unassuming, he felt he was just an anonymous engineer in a big group of engineers.

One day while browsing in the public library, Kevin delved into the topic of visual aids, remembering how inadequate visuals usually were at technical conferences. His interest grew when he discovered that no single book or article covered the subject adequately, so he researched the topic, took notes, outlined and wrote an article on the proper use of visual aids. It was accepted by a periodical, and he received a check and a copy of the periodical a short time later.

The next day Kevin's supervisor called him into his office and complimented him on his article. That afternoon the vice president of marketing called Kevin and sought his permission to reprint the article and distribute it to all of their employees.

During the next two weeks Kevin received calls from four different companies asking his permission to reprint the article and distribute it to all their employees.

Suddenly the anonymous engineer became known, not only in his own company, but throughout his profession.

This could happen to you once you break out of your indecision, take the plunge, and write an article for publication. It can change your professional and personal life, give you more confidence, and reward you in many other ways. Most of all it gives you a special inner satisfaction of self-accomplishment that few in your profession earn.

Writing is not a dull, unnecessary task to be endured. Writing is an important part of your work that can do much to advance your professional career. If you wish to be really successful, you must learn to write well. Writing is often the only tangible result of your efforts. Engineering/scientific writing is a skill to be learned, a professional tool that is as important as your technical knowledge and experience. Technical people who write well are readily known and recognized by their supervisors and contemporaries. Otherwise their ideas, no matter how good, may be ignored and overlooked. And being ignored and overlooked for one's honest efforts is enough to discourage any creative person. Recognition, however, leads to advancement, new assignments, and salary increases.

BUT FEW ARE PUBLISHED

Only 1 percent of all degreed engineers ever have an article published. This book shows you how to join this elite group of professionals.

Writing of technical matters always pays off. When you've published a number of good articles, a few scientific papers, or a solid book, it can make the difference between a routine career and an outstanding one. As an author, you're certain to gain professional recognition.

Studies have shown that most of the knowledge scientists and engineers learned in college becomes obsolete only four to seven years after they graduate, unless they continue their education. Obviously most professionals can't afford to take a sabbatical every few years to learn about the new developments in their profession, so they must learn about them by reading periodicals and books and by attending seminars. In some technical fields, as much as one-third of a workday must be devoted to keeping up with the ever-increasing knowledge of their specialty. Thus, all engineers and scientists incur an increasing obligation to write well so they can tell of their own developments and pass on this information to their peers.

If you're an engineer, a scientist, a technician, or an executive and have long wanted to have a scientific or technical article published, but you don't know where to begin or are afraid to start, this book will show you the direction. Techniques in this text have been developed to give you the impetus and confidence to direct you through all the necessary steps, in the proper order, to accomplish this all-important, career-advancing task.

NEW WRITING FORMULAS AND TECHNIQUES

The writing formulas and techniques described in this book will show you how to outline, organize, research, illustrate, and write all kinds of scientific and technical articles — and get them published! Perhaps you have some technical innovations or ideas you want to share with others in your profession. This book shows you how to make a national, even a worldwide audience aware of your accomplishments.

If you're in research or teaching, you know the value of being published for obtaining grants, tenure, and promotions. The more you publish, the more impressive are your credentials. In some academic institutions, you publish or perish.

EARN MONEY FROM WRITING

If you would like to earn money by selling your technical knowledge, this book will teach you not only how to write, but also how to market your articles. There are more than 6,000 technical, business, academic, and trade publications in the United States. These publications serve every conceivable occupation; they publish hundreds of thousands of technical articles and papers each year. It is indeed a fantastic, hungry, growing market.

If you're ready for that ultimate in technical writing — a book — all of the principles taught in this manual can be directly applied to that prestigious and profitable undertaking. A series of well-written articles that is accepted within the scientific and engineering communities is a pathway leading to a full-length book.

WHY WRITING IS DIFFICULT

Two of the biggest problems most writers face are:

1. What to write first.
2. What to write next.

The big, blank, white page staring up at you can paralyze even the most confident writer. What to write first often discourages many potential writers. Even though they bravely break down this initial barrier and write something down, they panic again about what to write next, and next, and next. They wonder if they're including all of the information, in the proper order, and. . . .

In spite of these psychological barriers, people's creative minds literally burst with thousands of ideas. These ideas come to their minds in a random order, however — a cacophony of creativity, seemingly unrelated to each other, but inescapably dictated by the complex interrelationships of each writer's background, loves, hates, experiences of all types, as well as cultural, ethnic, and educational backgrounds.

So, the problem is not how to come up with ideas, but how to organize ideas, how to use them effectively, and how to write them down in some meaningful order.

That's one of the main thrusts of this book: to show you how to organize your ideas so they form the basis of a cogent, lucid, and logical presentation.

Organization is a vital key to
unlocking your ideas and putting them in writing

THE PRECEDENT SORT

Order is essential to our lives. During every waking moment our minds are busy trying to draw order from confusion, trying to impose some pattern on the flood of stimuli flowing toward our senses. Without order, we live in chaos. In the Bible, Job speaks of death, not only as a land of darkness, but as a "land without any order." And order is essential when you read these sentences. Without organization you cannot comprehend what is written. It becomes a meaningless jumble. Order is an absolute essential to our comprehension.

Most people hate to organize simply because no logical method has been available. Yet they acknowledge that a comprehensive outline is essential in order to write a clear, concise article. Proper organization provides focus, direction, and impact for your writing. Facts don't usually speak for themselves; they speak only when related to the main topic of your article. Because the significance of facts grows out of their relationship to your topic, a logical organization is vital for readers to properly interpret and understand what is written.

What you need is a logical, painless method to organize your material, a method as easy to apply as the mathematical formulas that govern so much of engineering and science. And that's what you're going to learn in this book: how to write formulas.

You are not going to be bored by illogical, antiquated, hard-to-use methods containing Roman numerals, indentation, subordination, incantation, and other outdated ideas replete with confusion that were force-fed to you in high school and college. Rather you're going to learn a modern organizing method based on up-to-date computer technology, called the *precedent sort*. The precedent sort is easy to use, effective, and universal. It's like putting numbers in a simple formula and voila: the outline emerges!

Starting with a comprehensive outline, you will find it much easier to write and convert your ideas, your topics, into words and to string them all together to compose sentences and paragraphs until, before you realize it, you've written an entire article and you're ready to start your next one.

PEER FEAR

Another problem facing would-be writers is peer fear. The fear of writing a substandard article has deterred many engineers and scientists from being published. But you can overcome that barrier by following the procedures taught in this book to research, organize, and write your first article. You will learn what it takes to produce a publishable article and so gain confidence. Once you are published, your technical expertise will become known, you will become a recognized expert in your specific field, and others will seek your counsel. You'll meet new people, establish new contacts, and absorb more knowledge in talking with others.

BUT I DON'T KNOW HOW TO WRITE!

This book is all about ridding you of that mental block, showing you how writing can be easy. This text takes much of the mystery out of writing of things technical. Included in this practical manual are a compilation of pragmatic writing techniques, "tricks of the trade," tips that I have developed, distilled, tuned, and refined in over 25 years of scientific and engineering writing. My results were three nonfiction books; over thirty articles in national magazines; hundreds of technical manuals, reports, proposals; and thousands of letters and memos.

And don't think you need an enormous vocabulary to write. Studies have shown that a vocabulary of a mere 1,000 words covers about 85 percent of a writer's requirements on ordinary subjects. When you are writing on a specific technical subject, these basic words will, of course, be supplemented by the special words of your profession.

Most books on writing spend a considerable amount of time on grammar, but devote very little to the practical aspect of technical writing: that vital aspect of how to communicate ideas to others.

This text is different. You'll learn how to communicate effectively with words, using what I call *invisible writing techniques*. You'll learn writing formulas — guidelines that will help you create and organize your ideas. Then you'll learn techniques that will help you convert your organized notes into sentences, paragraphs, and a completed article in which you successfully communicate your ideas without the reader realizing how you accomplished it — a clever, silent seduction of sorts.

With the confidence and abilities you'll learn from this book, you'll overcome the unfortunate hang-ups and fears instilled by your high school and college English teachers. You'll find that writing of things technical can be easy, fun, and one of the most rewarding and stimulating activities you can participate in.

BEING A WRITER HAS ITS ADVANTAGES

Once you start writing using the techniques in this book and begin putting your ideas down on paper, you'll find that it has helped you in many ways:

- Writing your ideas down on paper will help clarify and organize your thoughts.
- You'll learn more about any subject you choose when you research, outline, and write your article.
- Once they are written down in cold type, you can study and evaluate your ideas with a critical, perceptive eye.
- Often your written documents are all that upper management sees of your work. Write well and you'll be recognized and rewarded for it.
- You're contributing needed knowledge to your field — helping others to learn.
- You might find, like so many people, that you can express your thoughts better in writing than in speaking. When you write, you can revise your material again and again until you get it just right. You can't do that when you speak. Once said, it's said, or as the Russian proverb tells us:

  A spoken word is not a sparrow.
  Once it flies out, you can't catch it.

- You'll soon discover it's easy to compose other types of important communications, such as simple memos, letters, reports, and articles. The techniques revealed in this book work for all types of writing.
- You'll help your company's business and improve your own chances of promotion.
- A list of publications looks impressive on your resumé.
- The technical person who writes well is looked upon as a leader in his or her profession.

- At a technical conference you may reach, at most, 100 people. In a technical periodical, you can reach tens of thousands of people.
- You'll earn an inner satisfaction, a pride in what you've accomplished. This, many feel, is the most important reward of all.

THE IMPORTANCE OF CLEAR WRITING

Most people recognize the names of Charles Darwin and Albert Einstein. Not only were their discoveries important; these men also were outstanding writers, able to communicate about complex things in simple terms that both scientists and laypeople could understand. Many brilliant scientists have died in obscurity, their discoveries unknown because they wrote leaden prose, foggy sentences, disorganized dissertations that no one could comprehend. It was almost as if they spoke a foreign language that no one understood. Their ideas had to be rediscovered by others because they did not properly document and report their original discoveries.

An example of leaden prose is the average Ph.D. thesis about which J. Frank Dobie wrote in his *A Texan in England*:

> The average Ph.D. thesis is nothing but a transference of bones from one graveyard to another.

Much technical writing will never be read because it is nothing but leaden prose, a rearrangement of old bones.

To illustrate how leaden a bad writer's prose can become, even in everyday life, read this:

> "In accordance with our previous agreement, it is hereby requested that you cease delivery from the period beginning Tuesday, May 19, up to, but not including Thursday, May 21, and thereafter, until further written notice is received from the undersigned."

In simple terms, this is a note to the milkman, written by a government bureaucrat, that should have read:

> "Please skip Wednesday."

Which version would you prefer to write, and to read?

An excellent example of effective brevity that was composed over a thousand years ago is the best trip report ever written:

> "Veni, vidi, vici."

This trip report by Julius Caesar translates to:

> "I came, I saw, I conquered."

So, to put it graphically, this book teaches you how to take the lead out of your prose.

STEPS TO FOLLOW

This book is organized according to the steps you would logically follow in writing an article as diagrammed in FIG. 1-1. Most of your time will be spent on research, organization, writing, and revision. Experts recommend devoting the following percentages of your time to these major functions:

40%   Research and organization
30%   Writing
30%   Revision

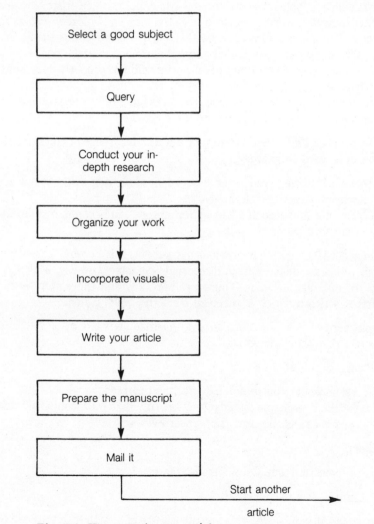

Fig. 1-1. How to write an article.

This breakdown illustrates the importance of thorough research and organization before you start writing. If you begin to write before you organize your material, it's like pouring concrete before setting up the forms. To tear apart the formless sentences and paragraphs you have written is exhausting and exasperating labor.

These percentages obviously can vary, depending on the subject you've chosen. However, if your writing takes up more than one-third of your total time, you're not spending enough time in researching and organizing, and your revision will be laborious and require excessive time.

EXERCISES

1. List 10 subjects you'd like to research, write about, and publish in a technical periodical. · *What's the difference btwn ISDN + B-ISDN?* · *What's the relationship btwn ATM + SONET* · *How is UNIX being migrated into the mini + PC + LAN environment, and why* · *Is downsizing really cost-effective?*

2. List 10 technical periodicals you'd like to write an article for. · *Datapro* · *Telecommunications* · *CommWk* · *Netwk World*

3. Why do you want to learn to write better?

4. Find an example of well-written technical material and a contrasting example of poorly written technical material. Check which of the following characteristics separate them into good and bad.

   ✓Organization
   ✓Long sentences   *Never use abbreviations until you've*
   ✓Long paragraphs   *spelled out completely 1st*
   ✓Long, complex words
   ✓Excessive use of jargon
   ✓Subject matter

5. Choose a highly technical article about a subject in your field of specialty. Select a half dozen consecutive paragraphs and rewrite them for a layperson. Translate the technical jargon so it can be understood by nontechnical readers.

   Note: I believe that today's reader is desperate for synopses + for primers/tutorials. Anything that will allow them to learn more in less time with no hype.

2

How to Select a
Publishable Subject

*Curiosity is one of the permanent and certain
characteristics of a vigorous mind.*

Samuel Johnson

Whether you have already chosen a subject or whether you need to select one, the material in this chapter will benefit you because the single most important decision you are going to make is to choose a subject. Select a good subject and your article will blossom and be easy to write. Select an inappropriate subject, and writing it will be a drudge; publishing it will be impossible.

This chapter will take you, step by step as diagrammed in FIG. 2-1 from defining what makes a good subject, through the final steps of selecting a subject and creating a working title.

START NOW

Don't wait until you've finished this chapter before you start thinking of subjects. As you read through this chapter, and as you go through your daily activities, keep a pen or pencil and paper beside you to jot down a number of possible topics you're going to consider. Keep a folder to file them in. Call this your Idea File.

Topics in this chapter will jog your memory and help you create dozens of suitable topics for articles. Generally the best subject to consider first is your work, something you know well, or a topic that interests you. It's exclusively yours and you should have enthusiasm for it. But don't limit your subjects to

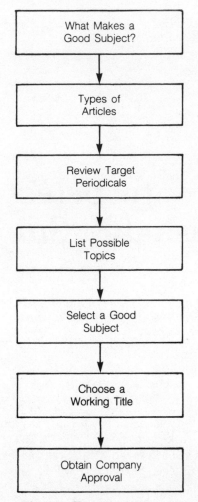

Fig. 2-1. How to select a good subject.

your work. Consider a wide variety of topics — let your imagination roam free and unfettered.

Creativity

Writing is a creative art. Scientific tests regard "facility in writing" as a basic index of creative aptitude. As a writer you are not simply going to copy down phrases, sentences, and ideas that others have already composed. Instead, you are going to review and evaluate what other writers have done with similar topics to see how their writings can influence, enhance, and supplement what you are creating.

Finally, you'll organize all of the information into a comprehensive whole, then write and revise your article until you are satisfied with it.

This process begins with that indispensable ingredient that exists in all people, in varying degrees:

creativity.

Imitate

Because you learn by imitation, you can enhance your creativity by consciously cultivating some of the attributes that characterize creativity. A creative person is:

- Unconventional
- Original
- Curious
- Sensitive
- Dissatisfied
- Open Minded
- Image Creating
- Fluent
- Persistent/Motivated
- In possession of a sense of humor

This might seem like a large number of characteristics to have, but most people possess these attributes to varying degrees. Creativity was first manifested in us when we were children — always curious, always asking questions, always trying new things. But this creativity was often stifled by parents and teachers who judged us harshly and impatiently, telling us not to ask such silly questions or do such stupid things.

Undoubtedly some of these cautions served a useful purpose, but in many instances it would have been better if we had learned the consequences ourselves.

As adults we became so conditioned to accept the status quo that we became afraid to suggest improvements for fear that our peers would judge us, laugh at us, or poke fun at us. There are thousands of examples of creative people being laughed at, from Noah to Fulton to modern-day researchers. But these creative people endured the criticism and carried their ideas to fruition.

That creative urge is still there in all of us, lying dormant, waiting to be nourished, eager to be used again. Curiosity is like a muscle: it must be exercised to be of full use.

And now you are going to have an excellent opportunity to exercise your "creativity muscles" and put them to work by writing an article that will surprise even you by its excellence. Review the characteristics that creative people possess. Ponder these priorities; practice and cultivate them consciously so they can become ingrained in you and become a part of your subconscious. Practice writing down ideas and thoughts, no matter how outlandish they may seem. Only you will be reading them. Ideas feed on ideas, so what might have seemed like a totally ridiculous idea can give birth to that great, revolutionary idea that can change your life.

Unconventional Creative people employ unusual problem-solving techniques and give uncommon answers to questions in the process of generating unusual solutions to problems. They balk at pressure to conform to the norm. These individuals are not afraid to try something different, to take a chance.

A creative person does not fit into a mold. The creative person is not necessarily the one who has the best memory. Studies have shown that people with exceptional memories are not creative, but imitative. As an example, an idiot savant can have a photographic memory, but is in no way creative.

The basic definition of creativity is the ability to come up with something new, something different. So a truly creative person is one who thinks differently, who sees objects and situations in a different light.

Original Creative people are always searching for something new and are not hampered by stereotyped solutions. Creative people can analyze existing systems, see room for improvement, and combine unusual connections and combinations to arrive at a solution. Creative people are receptive to new ideas and can rearrange basic elements in new combinations to form a new whole.

Sensitive Responding strongly to their senses of hearing, touch, taste, sensitivity to color, shapes, and textures also distinguish the creative individual. They have a greater feel for things and events that surround them. They visualize concepts better than most people.

Curious A predominant characteristic of creative people is that they have never totally lost their childhood wonder, the curiosity of their earlier years. They become intrigued by problems that puzzle them, and ponder how these problems can be solved. They also wonder about things, events, and processes. What caused them? Can the results be changed? Can it be done in a different and better way?

Dissatisfied Creative people are dissatisfied with the way things are —dissatisfied enough to change them. They don't accept situations as being inevitable, unchangeable. These people are restless, yet they cherish the opportunity to relax and switch their minds over to something different. This discontentment does not disturb them to the point of rejection, but to the point of wanting to improve it. They know there are better ways to accomplish things.

Open Minded A prime requirement of creativity is the ability to react to stimuli without prejudice, without shutting data out. Creative people open their minds to all ideas, are not bothered by the NIH (not invented here — If I didn't invent it, it's no good) factor, which prejudices so many decisions. An

I can do this!

ability to restrain critical judgment during the creative or idea-generation process, and openness to all approaches distinguishes the creative person. Don't toss out ideas before they can be considered, evaluated, and tested. Don't believe that the obvious way is the best way. Be receptive to unusual ideas. Creative people have a talent for ambiguity and can work in situations where no clear direction exists. They can find their way in the dark without a candle.

Image Creator   Creative people think in images, not words. They tend to be daydreamers and can visualize an idea taking shape. They believe it's best to solve most of the problem with images, even though their initial imagery may be foggy and ill-defined. They resist the temptation to articulate, to reduce their solution to words too soon, since words can put restrictions around an incomplete solution.

Fluent   Possessing fluency, creative individuals can generate fifty uses for a paper clip. They are fluent not only in their own field, but in many related fields. Fertility of ideas also characterizes these individuals. They can generate a large number of ideas in a short time. Ideas spawn other ideas, loosening the gates of imagination so that a large number of potential solutions emerge. As a result of this chaining effect, an outstanding solution usually will result.

Flexibility   The ability to adapt and adjust to new and changing situations is also important. Creative individuals can abandon old ways of thinking and initiate different directions, different problem-solving approaches. They can generate a number of different kinds of ideas and are able to break away from conventional methods of solution.

Persistent/Motivated   Creative people don't work an eight-hour day; they do not function to a schedule. They tend to be totally immersed in their work and fiercely determined to succeed. Not overly discouraged by failure, they realize that some failures are necessary in learning to walk the path to finish the work they have started. They are driven and persistent. These are probably their most important and strongest characteristics. More energetic, they have a strong desire to create, and they welcome confrontation. They are not discouraged by the unknown; they accept it as a challenge. Creative people have a huge capacity to take pains. They are stubborn—determined to see their project through to completion.

Sense of Humor   As George Orwell said, *"Every joke is a tiny act of rebellion."* This is characterized in creative individuals by their ability to laugh at life's foibles. Some experts claim that computers will never be able to do creative thinking because computers have no sense of humor. It is a healthy sense of humor that creative people have, not a degrading, cynical one.

KINDS OF ARTICLES

Some of the basic kinds of articles that appear in technical periodicals are described here. These are not strict divisions, and often an article can cover more than one of these kinds. This list is not in any way to be considered restrictive. Use it to jog your memory in your search for topics.

Design Procedures/Ideas　This kind of article shows your reader how to design a device, plant, or product, or how to devise a procedure. It can describe the requirements to be met, alternative methods to meet them, design steps, how to check the results, special considerations, future modifications, and projected results.

Process Description　A specific process is detailed in this kind of article. The process might be semimanual or automated. The raw materials and equipment required are covered, along with reliability, efficiency, and step-by-step operating procedures, including time, temperature, and pressure. Such articles also compare the advantages and disadvantages of the process with competitive processes.

Technical Descriptions　In this kind of article, a description is given of the general manner in which some device performs its functions and might include a general description of the device, its operating theory, design considerations, lists of major parts, reliability, and related factors.

New Products　Consider this popular kind of article carefully so that it does not become a thinly disguised product promotion piece. Stress the general use of the device, process, or product; construction data; methods of using, operating, and applying it in various situations; and anticipated results.

Management Techniques　A wide variety of topics are available for this field. Many deal with the methods and means of supervising, scheduling, and motivating employees. Since people form the most important asset in any industry, this kind of article, if it is written to cover very general management problems, is always in high demand.

Computerization　The use of computers is revolutionizing all industries, so articles on how computers are being used are always needed. The list of potential applications is virtually unlimited.

Product Surveys　This popular topic provides an unbiased comparison of competing companies' products, processes, and devices. Write this one carefully, or you might alienate one or more companies that advertise in the periodical.

Plant Descriptions　This kind of article covers the functions of the plant, considerations in location and size, equipment installed, plant capacity, plant flow, operating details, personnel facilities, and any unusual design features, such as degree of automation, energy management, and environmental design. This kind of article usually proceeds from the outside in, describing the overall forest before detailing the trees.

Mathematical/Computer/Graphical Solutions　A need always exists to provide better, easier, and more accurate solutions to a wide range of design problems. Solutions can be purely mathematical, by computer, or by graphical means. Covered are calculation procedures, equations, assumptions, limitations, and one or two examples of the procedure.

How Tos The scope of this kind of article is virtually unlimited. A how-to article can show you most anything related to any aspect of your profession. They can cover how to schedule a project, how to write specifications, how to save time in a process, all the way to a book on how to write better technical articles. How-to articles will always be in high demand.

Miscellaneous The miscellaneous category covers a wide range of possible topics, such as:

- Opinion articles — Opinion articles are difficult to sell because most technical people like to separate personal opinions from their work.
- Plans for future projects — Before projects start, or as a progress report after a project is underway.
- Reporting what others have done — You might want to write about some special project underway at your company that others are working on. Often your co-workers are unable or unequipped to write about their accomplishments and need a technical reporter to write about it for them.
- Humor — Unfortunately the demand for humor articles is not too great, giving some credence to the stereotyped serious periodical editors. I once wrote a humorous satire claiming the word man-month was a sexist phrase. The article was published in a national magazine and I received many letters and calls from entertained readers and an invitation to speak at an electronics convention. However, the publisher apparently wasn't impressed and wouldn't consider any more humorous articles for his staid technical periodical.

JOT DOWN YOUR IDEAS

As you read through the various sections of this chapter, you might find that certain topics you've listed in your idea file are not suitable for a variety of reasons. If so, cross them out. But don't be too critical at this point; save the final culling out for later.

As you read and as you conduct your daily activities, continue to write down any more topics that occur to you. By the time you reach the end of this chapter, your list should be extensive and you should be ready to evaluate the topics and select the ones for your first few articles. Next, let's see what criteria a good subject must meet.

Benefit Your Reader

A primary requirement for all of your articles is that your topic benefits your readers. If it does not, it simply will not be published. Most technical articles are written to inform readers of new and future developments: how things work or how to accomplish certain design tasks or processes. Technical articles are written *by* technical professionals *for* technical professionals. All topics must be slanted so technical people can profit from reading them. A few articles are written to inform, but most are published to educate readers, to teach them more about their profession. And the more readers you can attract, the greater the likelihood your article will be published.

Always keep this question foremost in your mind when you select a subject:

"How will my article benefit my readers?"

If you can't come up with a solid answer, select another subject.

Be Topical

Technical periodicals usually cover topical subjects; that is, topics that are currently in use, recently developed, etc.

Technical books, however, cover topics that are of a more archival or long-lasting interest — subjects that are not out of date by the time the book is published. In recent years the division between the two has blurred somewhat; however, the content of most technical periodicals is still principally topical.

The reasons for this difference in coverage are twofold: First, a periodical can be written, printed, and distributed to a list of subscribers in less than a month's time, before the new gets old. A book, however, often takes a year or more from start to finish. Secondly, and perhaps more importantly, periodicals derive most of their revenue from advertising. Many technical periodicals are given free to the reader, the publishing and distribution costs being covered by the advertisers. Technical periodicals must attract new readers to every issue so their subscribers will keep reading, and hopefully responding to, the latest ads.

However, a book earns its revenue from the sale of the book alone. It is purchased solely for its contents. Readers expect topical subjects in technical periodicals, so if you want your articles to be published, *be topical*.

Be Unique

Uniqueness covers the superlatives of the professions:

- Newest
- Biggest
- Smallest
- Fastest
- Cheapest

If your topic covers a product, a development, or a process that falls under one or more of these categories, you have a built-in readership because one of the main functions of technical periodicals is to inform technical personnel about how to make things smaller, less expensive, faster, etc. You can fulfill this need if you discuss such a characteristic. If applicable, use one of these superlatives in the title of your article.

Still, uniqueness isn't an absolute requirement. If you have a new slant, a new approach to what has been published before, it's also an excellent candidate for an article.

Interest a Wide Readership

This might seem like stating the obvious, but so many people become so involved in their work that they don blinders and completely ignore this simple fact: the broader the appeal of your topic, the more likely that your article will be published.

Don't write about the design you completed using an A29BXT4 integrated circuit in your automatic bean counter for the accounting department. Broaden the appeal of your design; consider a host of other general counting applications, such as counting lima beans, jelly beans, cars, people, railroad cars, etc.

When you designed your counter, you might have concentrated on designing for an accounting bean. But unless your design is useful for dozens of other applications, few will read your article. Your readers are primarily interested in their own application, not yours.

Something You Know or Can Research

Ideally your subject should be something you know a lot about and can write about with authority. Your readers will recognize this authority and respect you for it. If you're the only person to your knowledge working on some subject and it has wide appeal, you greatly enhance your chances of publication.

However, you can also write about topics you know little or nothing about if you follow these words of advice from chapter 4:

Research the literature

If you find a topic that intrigues you and that you'd like to learn more about (your interest is vital in this approach), you can devote the necessary time to research your subject. When you thoroughly research a topic you learn a lot about it and become an expert through osmosis. Then you can write about it with authority, even though most of your knowledge was acquired from the writings of others. This is research, not plagiarism; because you're reading, critically evaluating, reorganizing, and updating the work of many others, incorporating your own ideas and slant on the topic.

To illustrate, I searched for a suitable topic for a technical article, but my work at the time wasn't a suitable topic. So, after doing a little preliminary research, I chose "Visual Aids for Technical Talks" as my topic. It was a subject that I felt had not been adequately covered in periodicals and one that I felt most readers could benefit from. Sooner or later most technical people have to give a speech requiring visual aids, whether it's a proposal presentation, a request for a grant, a progress report to management, or a technical talk.

Even though I had given a few technical talks at conferences in my professional career, I could in no way consider myself an expert on visual aids. So I went to work, thoroughly researched the subject, reviewed books and periodicals on the topic, took notes, etc.

Then, using the techniques described in this book, I wrote a query letter, received a positive response, and wrote the article. It was accepted the first time out. After it was published, I received a number of letters and calls praising the article and its practical content. Some people asked permission to reprint it as a guideline for use by their companies. In spite of the mostly library-acquired expertise on the subject, I hadn't originated any new information. I had merely researched, adapted, organized, and rewritten existing information, adding my own ideas and slant for the article.

You can do the same and discover dozens of topics you're intrigued by. Research, evaluate, organize, query, and write about them and you can also become a published expert.

And don't think you need anything of earthshaking significance to write about. Most articles you see in technical journals barely nudge the Richter scale. They are simply everyday topics that the author has researched and written about competently, in a manner that others can read, understand, and profit from.

The Periodicals You Want to Write For

This should also be so obvious that it shouldn't have to be mentioned, yet technical editors are constantly plagued with articles that are totally unsuited to the type of material they publish. It should be obvious that you don't send a technical article on the design of lasers to *Readers' Digest,* nor one on getting along with your mother-in-law to *Scientific American* (unless she's a nuclear physicist). Make sure there is a market for your article or you are wasting a lot of valuable time.

REVIEW TARGET PERIODICALS

Now that you have a pretty good idea of the kinds of topics that make for good copy, the next step is to review some of the periodicals in which you'd like to be published. A review form is provided at the end of this chapter to record this information. Xerox it so you'll have a number of blank forms to fill out when you're evaluating periodicals.

First, don't choose only one periodical to write for because the odds might be too much against you. The idiosyncracies of the editors, the advertisers, and the number of the editor's relatives on the staff all work against you. Select at least three or four periodicals in your field of expertise, periodicals you like to read. (That's a good standard: if you like to read their articles, there is a good chance that you can be published in them.)

You can probably find back issues of periodicals in your company or public library, or from your local pack-rat co-worker who saves old issues of everything. A representative list of periodicals is included in appendix C.

Even though its primary purpose is to list advertising rates. *Business Publications Rates and Data,* published by Standard Rate and Data Service of Skokie, Illinois, is an excellent source for checking on periodicals that carry advertising. SRDS lists the following:

- The publisher and editorial staff
- The publisher's editorial profile — the types of articles published
- Circulation — paid and nonpaid
- Business analysis of circulation — types of products, services covered
- Breakdown of the readership by job title

Usually a periodical's last issue of each year lists all of the articles published in that year. Check one of these last issues and review the titles. You'll get a quick, capsule view of what types of topics are popular in the periodicals of your choice. And you'll also get a taste for the kind of titles used in the periodicals. You might even want to make your own checklist to see how many of each kind of article discussed earlier in this chapter were published in each periodical.

Go through Back Issues

Next, go through three or four back issues of the periodicals you want to write for, page by page.

Although it may seem like a lot of work at first to review the periodicals you'd like to be published in, this effort will help you pick the right subject, the best title, the correct style, and will practically guarantee your article's acceptance. Like in most of life's endeavors, the more effort you put into it, the more rewarding the results.

What to Look for

Check the cover first. Are photographs and illustrations used? Are titles of some of the key articles listed on the front cover?

Review the table of contents. What types of articles were published, were they how-tos, informational, product surveys, design procedures. What is the average title length? What types of titles are used?

The next important point of contact should be the *masthead* of the periodical that should be displayed on one of the first few pages. The masthead will list the editor and the editorial and advertising staffs. It will tell you who to address your query letter to. Usually this will be the editor; however, if the periodical has special department editors; address your query letter to that special editor. Check the masthead against the table of contents to see how many of the articles were written by a staff member. If most articles are staff-written, your chances of publication in that periodical are poor. Concentrate on those periodicals that publish mostly free-lanced articles.

If you find an article written by the editor, read it carefully. It will give you a good idea of the type, slant, structure, language, and subject matter the editor favors.

Read the letters to the editor. They'll indicate how earlier articles were received and which drew the most positive or negative responses. If the editor has written an editorial, read it carefully. It might provide additional information on the types of articles the periodical is seeking.

Review the Ads

It's the advertising revenue that pays for the free (or low-cost) subscriptions to the periodical. And the advertisers, who want their advertising revenue invested wisely, are experts at pinpointing the types of people who read and respond to their ads by obtaining readership profiles from the periodicals.

A careful analysis of a periodical's ads can show you:

- What the typical reader of the periodical is like (interests, education, profession).
- The types of subjects the editor is seeking, since the articles' contents frequently complement the ads.

Often there is a close correlation between the types of articles that are published and the ads. Advertising agencies prepare lists of articles that will be featured in forthcoming issues and use this promotional material to interest companies to advertise in these specific issues. So you see the importance of studying the ads and using this information in selecting a topic for your article.

To analyze the ads, determine:

- What types of products are advertised: components, products, services, systems, books?
- What type of people and products are depicted in the ads: young, old,

professionals, blue collar? How are they dressed? What are they doing in the ad?

And don't forget the mail-order ads that are usually in the back section of the periodical. They also give some clues as to the types of products readers are interested in.

One important piece of advice about ads: if you want your article published, never antagonize an advertiser in your article! The reason should be obvious.

Study the Visuals

Review the visuals. Are photographs used? Black and white or color? What kind of captions are used on the visuals? Are line drawings, cartoons, graphs, and curves used? Is the art complex? How many visuals per article?

If photographs are used, you should determine in your query if the publisher prefers transparencies, 8-X-10-inch glossies, or contact sheets. Polaroids or low-quality prints are generally not acceptable.

Style

Study the articles to see what styles the writers use. What type of vocabulary is utilized — long words or short words, considerable jargon or none?

What is the sentence length? Are a variety of sentence lengths used? Are verbs active or passive? Are descriptive words used? Are contractions used? Is first person used: I, you, or they?

One method of determining the educational level of readers is through Gunning's Fog Index. To compute it:

1. Take a sample of 100 words of continuous writing and measure the sample with a ruler to see how many column inches the 100 words take up. Make a note of this number.
2. Count the number of words per sentence in the 100-word sample, but treat independent clauses (the parts of a sentence separated by a semicolon) as separate sentences. Calculate the average sentence length.
3. Count the number of words of three or more syllables in the 100-word sample. Don't count capitalized words, easy combinations like book-keeper, or verbs with three syllables ending in -es or -ed (for example, *edited*).
4. Add the average sentence length in Step 2 to the number of polysyllable words per 100 words. Multiply this figure by 0.4.
5. Pick three or four more random samples of continuous writing and

repeat this procedure. Average the result. This will give you the number of years of schooling a reader would need to read the article with ease and understanding.

Following are examples of Gunning's Fog Index for the world's best seller and for some popular magazines:

The Bible	6 – 7
Reader's Digest	8
Time	10
Atlantic Monthly	12

CHOOSE A TOPIC

By now you should have an idea file, bulging with at least ten or twenty potential article topics. Next, sit down in a quiet place with a pen or pencil and read over your list. As you read, more new ideas and new slants will occur to you. Write them down, no matter how impractical they seem. You can weed out the bad ones later.

Visualize yourself in the reader's shoes. What would *you* like to see in the periodicals that you and your co-workers read? Write these ideas down, using just a few words to describe your topic.

Finally, talk to your co-workers. Ask them what they would like to read about, what topics haven't been covered, and what they like or dislike about articles in current periodicals. Note down these additional ideas.

Now that you have a long list of potential articles, it's time to narrow your list down to three: one primary topic and two backup topics just in case your primary topic doesn't work out. Carefully evaluate the topics and cross off the ones that do not meet the criteria established earlier in this chapter.

SELECT WORKING TITLES

With three topics selected, it's time to pick working titles for all three. The title is an extremely important part of your article — it's usually the first thing a reader sees. Your title will be read by thousands of people, but probably only a few hundred will read your entire article. If you create a just-right title when you write your query, your chances of being published will be much enhanced.

Because of the mountains of literature available, and with more being printed and circulated every month, your reader is pressed for time and will first skim your title to determine whether or not to read on. So your title must be good enough to hook your reader. Investing a lot of time and effort in creating a grabbing title will result in big dividends.

A descriptive title is also important so people using on-line data banks (see chapter 9) can key in the proper words to locate your article after it has been published.

When you performed your periodical survey, you probably noted how often certain words appeared in titles:

- How to
- New
- Unique
- Smallest
- Easy
- Fastest

A good title is actually a summary of an article, but it can't be too long—no more than six to eight words. Use the active tense in your title; avoid jargon. Use words that are familiar, specific, common, and short.

Types of Titles

A title that hooks gives the reader a little information and entices the reader on. Sometimes a title need do no more than pique a reader's curiosity—a curiosity that can only be satisfied by reading your entire article.

Four types of titles are commonly used Label, Question, Imperative, Statement.

A *label* title can consist of one or more words and is a name or label that describes the contents of your article; for example:

- Consulting
- Brainstorming
- Computer Gold

The label title is the type of title most often used.

A *question* title should lure the reader into the body of your article. For example:

- *Ideas: Where Do You Find Them?*
- *Is Your Chemical Process Foolproof?*

When properly used, question titles can titillate your readers enough so they must read your article to find the answer to the question posed.

An *imperative* title urges your reader to action and is often used in how-to articles:

- *Check Your Unit's Reliability Under Operating Conditions*
- *How to Design a Suspension Bridge*

This is also an excellent type of title to use for informative articles.

A *statement* title is a short sentence that summarizes your article; for example:

- *New Computer Techniques Improve Reliability of Communication*

When generating titles, avoid, if possible:

- Articles (the, an, a).
- Numbers and complex symbols. Technical titles should be informative, not mysterious.
- Long titles.
- Vagueness — Remember that your title might be the only information given on your article in abstracting journals and indexing services.
- Unnecessary words, such as: *Study of, Investigation of, A Final Report On.*
- Passive voice.

Titling Techniques

When people start to read a periodical, they might first look at the front cover, then the table of contents, then start flipping through the pages until a title or a visual catches their attention. If their interest is tweaked enough, they'll stop and begin to read your article. Your title or visuals have then accomplished one of their missions.

You can use a number of techniques to further enhance the saleability of your title, to make it stand up and be noticed, even in staid, conservative periodicals.

- Alliteration — *Computer Commuter*
- Parody — *What a Difference a Computer Makes*
- Play on Words — *Only Your QC Department Knows*
- Rhymes — *Spy in the Sky*
- Coined Words — *Euphemania*
- Paradox — *Spend Dollars and Save Millions*

Write down a number of possible titles for each article and select a working title that meets these criteria.

OBTAIN COMPANY APPROVAL

Most companies have regulations that their employees must obtain permission before they publish articles, even if they are going to write about something not connected with their work. Before you query publishers, check your company's policies on this matter. Can you submit material directly to publications, or must you go through your firm's public relations or advertising departments? Is it necessary to have the material screened by your legal department for patent or confidentiality conflicts?

PERIODICAL ANALYSIS

Periodical Title _____

Date (Mo/Yr) _____ How often Pub.? _____ Free or Sub _____

What's on cover? _____

Editor's name _____

Special Editors _____

Editorial subject _____

Table of Contents — No. of articles _____

How many are staff written? _____ Free lanced? _____

Ads _____

_____

_____

_____

_____

_____

_____

Comments _____

_____

_____

_____

Fig. 2-2. Periodical analysis form.

How to Select a Publishable Subject

ARTICLE ANALYSIS

(Use a separate sheet for each article)

Article title _____

Author _____

Column inches per 100 words _____ No. of inches _____

No. of words _____ Type of article _____

Avg. Sent. Length _____ Avg. Para Length _____

Visuals: Quantity _____ Types _____

Photos _____ Types _____

Special visuals or special effects _____

Comments _____

_____

_____

_____

_____

Fig. 2-3. Article analysis form.

EXERCISES

1. Generate at least twenty uses for a paper clip.

2. Narrow your list of subjects down to three: one prime and two backups.

3. Narrow the list of periodicals you'd like to be published in down to three.

4. Review your periodicals and narrow your list to one. Keep the other two as backups in case your first choice doesn't work out.

5. For the selected periodical, analyze at least two issues of the periodical using the analysis form in FIG. 2-2 at the end of this chapter.

6. In your selected periodical, analyze at least three articles using the form in FIG. 2-3.

3

Query

*He has half the deed done
who has made a beginning.*

Horace

Two basic approaches to selling an idea for publication are:

1. Query first, then write the article.
2. Write the article, then send in the completed article.

The first method is by far the most practical, since you need invest only a small amount of your time in preliminary research and in writing a query letter (a brief proposal) to sell your idea. If the editor likes your concept, he or she gives you a tentative commitment that the periodical will buy your article if it lives up to their expectations.

Thus, you sell your article before you write it. That's what most professional writers do, and that's how most editors prefer to work with writers.

IT'S IMPORTANT TO SELL YOUR EDITOR

Before we get too deep into developing a query, let's take a look at the most important individual you will be dealing with in your quest for publication: the editor. Whether this person is an editor-in-chief, a managing editor, an associate editor, or an assistant editor, this is the individual you have to sell your article to. True, the editor may have technical experts on the periodical's staff to "referee" or review your article, but the editor is going to make that first important decision whether to even consider your query, and then, whether to publish you.

More than any one individual, the editor knows what a reader wants because, if he or she doesn't satisfy the reader, the publication ceases to exist.

The Editor's Job

An editor's job is not an easy one. I know whereof I speak because I served as an editor on two technical journals for a number of years. Writing columns, prodding authors, editing papers, corresponding, attending meetings, scheduling, budgeting, attending conferences, and a dozen other activities fill an editor's day to overflowing. Still, it is a rewarding job to guide scientists and engineers through all of the steps necessary to achieve publication.

Technical editors come from both sides of the spectrum: technical and editorial. A constant debate exists over which source contributes the best editors. Although I came from the technical side (I worked as a scientist and as an engineer for many years), my experience has been that excellent editors can come from either profession. As usual in any profession, it's the individual's ability and determination that count, much more than a background or previous occupation.

Editor's don't have a monopoly on technical articles and must keep their pipeline full of articles in various stages of completion to fill their periodicals. An editor might have to deal with as many as five to ten times more authors than the periodical can possibly publish and may end up with four to nine authors with bruised egos for each article published.

Most editorial offices operate on a panic schedule. An enormous amount of work must be accomplished in a limited time in dealing with advertisers, printers, authors, mailing lists, computer typesetters, electronic paste-up, illustrations, etc. to produce a periodical that must always be completed and mailed on a fixed calendar schedule.

So, before you write your query, consider the harried editors you're writing to and make that query short, succinct, exciting, and enticing enough to make that editor smile and conclude, "I just have to publish this article."

Why Query First?

In addition to saving your and the editor's valuable time, a query will benefit you in the following ways:

1. You will be able to determine if the subject matter of your article is suitable for that particular publication. Requirements vary considerably from time to time, and a query is a quick way to find out the current editorial needs.

2. If the periodical is entirely staff written, a query will quickly establish this fact and save your effort.

3. You only have to wait a week or two for a reply to a query. However, a periodical may take two to three months, or more, to evaluate a completed article and render a decision.

4. The editor, in his or her response, will send you the periodical's editorial guidelines regarding length, format, types and number of illustrations desired, etc., so you'll have all of this information available before you invest your valuable time in researching and writing a full-length article.

5. You'll have established some rapport with an editor that will serve you well for this and future submittals. Editors prefer repeaters; they cultivate writers who can continue to deliver good articles.

6. It's the professional way of conducting your writing business.

MAGAZINES TO QUERY

A large number of representative periodicals are listed in Appendix B. Other sources have comprehensive lists of technical publications and technical associations in all fields of technical activity. They not only list the periodical, they also provide information about the topics the periodical covers. Some of these references are:

- *Encyclopedia of Associations*—Gale Research Agency
- *Scientific, Engineering, and Medical Societies Publications in Print*—James Kyed and James Matarazzo
- *Standard Periodical Directory*
- *Ulrich's International Periodical Dictionary*—R. R. Bowker

If your library doesn't have the periodicals you want to review, your friendly librarian can probably arrange to borrow a few back issues from other libraries.

SIMULTANEOUS QUERIES?

Suppose you're not certain that a specific periodical will be receptive to your query, so you decide to write to a number of different periodicals at the same time, with the same article proposal, on the chance that at least one will buy your idea.

In a word: don't!

It's possible that more than one editor might like your idea and ask for a completed article. Then you're in trouble. You'll have to renege on your original promise to some of the editors. Editors who have been turned down have long memories and may be biased against you for future articles.

Play it safe and submit a query letter to only one periodical at a time. If your proposal is rejected, immediately submit your query to another periodical . . . until you find a receptive editor.

VANITY PUBLISHING?

Instead of having the periodical pay for your article, some editors actually have the audacity to insist that the author pay to be published in their periodical or journal. These appropriately named, *vanity publishers* are an insult to the profession. Such ego publications are not worth the paper they're printed on.

If an editor replies to your query with, "We'd love to publish your article; however, our editorial policy requires that an author must pay $100 a page for the privilege of being published in our periodical . . . ," simply reply, "No thanks," and send your query to an editor who is buying, or who will at least publish it at no charge to you.

Stay away from vanity publishing. If you're going to invest your valuable time and effort in researching and writing an article, you shouldn't have to pay to have it published.

Some periodicals, often technical journals, operate on a limited budget. Because they are the official journals of nonprofit technical societies, understandably they can't pay for articles. If you feel that exposure in these periodicals will enhance your career, this type of periodical may be your best choice. But absolutely refuse to pay to be published!

TELEPHONE OR WRITTEN QUERY?

You can contact an editor by telephone or by letter. (In the future you may be able to send the editor your query by modem. See chapter 9.) If you believe that you can effectively sell your article on the telephone, do so.

Although a telephone query has the advantage of giving you a quick answer to your query, I believe it's best to send an editor a written query. A written query gives you time to plan what you're going to say, and to say it the way you want to. Also, it's less of an imposition on a busy editor's time if he or she receives a written query that can be read and pondered at leisure and discussed with co-workers before a decision is made. Often such decisions must be made during staff conferences, involving a number of decision makers, so a written query would serve best.

HOW MUCH PRELIMINARY RESEARCH IS NEEDED?

The amount of preliminary research you need to do before you write a query depends on how much you know about your subject before you query. If you're writing on a topic that you've been working on for some time, you need only search your brain and organize this information to write an effective query.

However, if you're writing on a subject about which your knowledge is limited, you're going to have to search the brains of others in enough depth to write a query that can convince an editor that you can write an authoritative, full-length article and deliver it on schedule. For this preliminary search, it's best to select one or more sources that are authoritative and recent to make sure your information is accurate and up to date.

Shortly I'll present a typical query letter so you can see what it contains and gauge how much preliminary research you must do to sell your article to an editor.

Before we cover a query letter, however, you're going to be introduced to the fabulous four-part formula. This simple, but powerful formula should form a basic part of everything you write, whether it's a short memo, a report, an article, a chapter in a book, whatever . . . Once learned, you'll find that the four-part formula simplifies your writing, as well as makes it more saleable, more attractive to a reader, and easier and more enjoyable to write.

FOUR-PART FORMULA

This old and well-proven formula for effective writing is based on the fundamental way in which people react. The four-part formula, described in Walter S. Campbell's excellent book, *Writing Non-fiction,* is valid for everything you write and is, simply:

- HEY!
- YOU!
- SEE?
- SO!

HEY! The first part of the formula requires that you catch the reader's attention by using some interesting phrases or attention-getting statement, such as:

"Sex is America's second favorite sport. Sex is also America's favorite sport."

YOU! For this part of the formula, convince the reader that what you said in your opening and what you are going to say in the rest of your article, affects, and is of interest to him or her. For example:

"Do you realize that if it weren't for sex, you wouldn't be here today?"

SEE? Now that you have your reader's attention, you are ready to write the *body,* or main part, of your article. In this section you present your facts, your story to hold your reader's interest. To show the reader how you prove your thesis, demonstrate the points you want to make and deliver the message you have to give, such as:

"Among the million and one advantages of sex are . . ."

SO!   In the final part of your article you convince your reader that he or she has profited from reading your communication. Leave your reader with some closing idea, thought, conclusion, summary, or a call to action of some kind:

"So, go out and get sexed. It'll clean the cobwebs out of your brain."

HOW TO USE THE FORMULA

Whether you realize it or not, you use the four-part formula many times a day. For example, suppose you see your co-worker, Clyde, walking down the hall at work:

"Clyde!" you call out. (HEY!)

Turning, Clyde walks over to you. "Yes?" he asks, frowning and tugging at his ear.

"How'd you like to go to the ball game tonight?" (YOU!)

"Who's playing?" Clyde responds, leaning against the wall.

"The Dodgers. I have an extra ticket. Pizzica is pitching and Sather is catching. Should be a super game!" (SEE!)

"Sounds great," Clyde smiles. "Okay."

"Pick you up at seven," you say. (SO!)

There! You've automatically used the four-part formula, without even having to think about it. It's a completely natural formula, based on the way people react to a situation. See how effective it is?

In commercial advertising, the four-part formula is also used extensively:

"Our Hartford automobile tires are triple-steel-belted (HEY!) to keep you (YOU!) safe on the roads. Tested on rough, washboard roads and high-speed freeways, Hartford tires show no wear after five million miles (SEE?). They're on sale at your local dealer. Buy a set today. (SO!)

And to illustrate the use of the four-part formula in writing a simple memo:

TO: Jennifer Johnson
FROM: Quentin Czarnecki
SUBJECT: Lab tests on Solvent X-23 (HEY!)

The lab tests you (YOU!) performed last Friday were not documented properly. We need your complete report so we can distribute copies to Research. (SEE?)

Please complete the report by Friday, January 13, or you'll be pulled off the project. (SO!)

When you start using the HEY! YOU! SEE? SO! formula consciously, you'll find that you will create more interesting writing that is effective, concise, and easier, almost formulalike, to write.

REQUIREMENTS OF A QUERY

A successful query must:

1. Be brief. A maximum of one page. Editors are busy people and are prejudiced against writers who can't state their case briefly.
2. Hook the editor. You must convince the editor that your idea is a good one that will benefit the readers.
3. Summarize your article. Give enough information so the editor can make an intelligent evaluation of your idea.
4. Establish why you are the person to write the article. Outline your unique qualifications.
5. Include your proposed title and specify the tentative length, number of illustrations, etc.
6. Assure the editor that you've cleared the article subject matter with your company management, if that is required where you work.
7. Tell the editor how long after go-ahead you'll deliver the completed article.
8. Be typed, single-spaced, neat, error-free.
9. Address the editor or editorial staff member by name. Don't use "Dear Sir" — it's a tipoff that you haven't even bothered to look at a copy of their periodical to obtain the correct addressee.
10. Ask for a copy of their writer's guidelines.
11. Enclose a self-addressed, stamped envelope (SASE) for the reply.

AT LAST, A SAMPLE QUERY

Since one of the fundamental adages about writing is "Write about what you know best," I'm going to illustrate in FIG. 3-1 with a sample query to sell an editor on publishing an article about helping authors organize their material when they write for technical periodicals.

Query

31 February 1999

Wayne Brent, Editor
Advanced Technology
1313 Idaho Street
Waterloo, TX 77079

Dear Mr. Brent,

1. Engineers and scientists, who love order and precision in things, hate to outline because so far no logical system has been devised for outlining. Yet, everyone acknowledges that a good outline is essential to write a good article. And you, as an editor can recognize when one of your authors has used an effective outline to produce an excellent article. (HEY! YOU!)

2. Scientific professionals love to use formulas. All they have to do is select the right formula, input the proper values, turn the crank, and come out with the correct answer. (HEY!)

3. So, why not a formula for outlining? (HEY!)

4. That's what I've been developing over the past few years, an easy-to-use formula for outlining technical documents, ranging from simple memos, to proposals, to technical articles . . . even to books. I've named this formula the precedent sort. It promises to modernize many aspects of writing of things technical. (SEE?)

5. I'd like to write an article for you which I've titled HOW TO ORGANIZE TECHNICAL WRITING. I can cover the topic in about 2,000 words, complete with examples of how to use the precedent sort to organize technical material. I'll use four or five illustrations. This will be the first time this technique will appear in public print, so I know your readers will be interested in, and profit from, this topic. (SEE?)

6. As for my credentials, I'm a full-time free-lance writer, working in both the technical and commercial fields. I have a B.S. in Engineering and have worked as an electronic engineer in both design and program management, so I am very familiar with engineering writing requirements. I've had three non-fiction books published and have authored over two hundred technical manuals, proposals, plus numerous technical articles published in national periodicals. (SEE?)

7. Please send me a copy of your writer's guidelines. I'm enclosing a SASE.

8. I can deliver a completed article one month after your go-ahead. I'll look forward to your reaction to this exclusive submission. (SO!)

Sincerely,

Harley Bjelland
Free-Lance Writer
P.O. Box 1776
Any City, MN 54321

Fig. 3-1. Query letter.

Use of the Four-Part Formula

The paragraphs in the figure have been numbered simply for reference. They shouldn't be numbered in your query.

This query follows the basic four-part formula. The HEY! and YOU! parts are somewhat intermingled, as they usually are. By the end of the third paragraph, the editor should be hooked, wondering if there really could be a formula for such a subjective task as outlining.

Paragraphs four through six cover the SEE?. The last paragraph covers the SO!, making a commitment and asking the editor to respond. Note that the editor is being informed that this is an exclusive submission, a requirement some periodicals insist on.

Analysis of Sample Query

Let's go back to the Requirements of a Query to see how the sample query meets the criteria established.

1. Brief. The query is about one page long.
2. Hook. The editor should be hooked, his or her interest aroused, by the end of the third paragraph.
4. Credentials are established in paragraphs four and six.
5. The title, length, and number of illustrations are proposed in paragraph five.
6. It's not necessary for me in this case to obtain company clearance since I'm a free-lance writer.
7. A copy of the periodical's Writer's Guidelines is requested and a SASE is enclosed.
8. The last paragraph makes a commitment to deliver in 1 month.
9. The letter is typed, single-spaced, neat, and free of errors.

SOME MECHANICS

To help with the preliminary planning of your article, here are some basic recommendations on length, etc.

The specific periodical you're aiming for usually has certain length requirements, although these are often flexible. Most articles range from 1,000 to 3,000 words; about 2,000 to 2,500 is the average.

One thousand words is about four pages of double-spaced, typewritten material, so your typed draft should be between four and 12 pages long.

As far as illustrations are concerned, use at least one or two for even the shortest article, and perhaps five or six (count tables as illustrations) for longer articles. For some periodicals you need not worry about supplying finished artwork since they usually redraw everything you submit to meet their own specifications. Some periodicals, often professional society journals, require camera-ready copy. Your editor should answer this question in response to your query. Photographs can be reduced by your publisher as required.

A typical three-column, 8½-by-11-inch page of a periodical contains about 1,000 words, without illustrations, so you can estimate the number of finished pages for your article. If you use the recommended two visuals per periodical page, they'll occupy about one-third of the printed page, leaving about 700 words per page when two visuals are used. You can expect to receive from $25 to $100 per page, up to as much as several hundred dollars if you have an outstanding article.

EDITOR'S RESPONSE

For a query, you should hear from an editor in two to four weeks. If you haven't heard after about a month, write a short, polite note to jog your editor's memory, asking if a decision has been made on your article.

Your editor's response can be one of three types:

1. Your query has done its job and you're given a go-ahead to write a full-length article.
2. Your proposed article is accepted conditionally; that is, the editor makes some suggestions for modifying your article (e.g., length, slant, etc.) before you submit your completed article.
3. Your proposal is rejected. Unfortunately this is a cold fact of the writing profession. Don't take it personally. There are many reasons for rejecting an article that have nothing to do with the quality of the article.

If you receive one of the first two responses, go to work: accomplish your in-depth research and write your article. If you receive the third response, before you get too depressed, quickly write another query to another periodical and mail it immediately! That's why I recommended that you have at least two back-up periodicals in mind in case your first choice doesn't work out. There's nothing like the renewed hope of being accepted the next time out to assuage the disappointment generated by a rejection.

EXERCISES

1. Narrow your list of subjects down to one.

2. Conduct enough preliminary research to draft your query.
3. Draft a query letter to the periodical of your choice.

4

How to Research the Literature

Knowledge is of two kinds.
We know a subject ourselves,
or we know where we can find
information upon it.

Samuel Johnson

To paraphrase Samuel Johnson a bit, the next best thing to knowledge is knowing where to acquire knowledge. And the best repositories of knowledge are libraries and commercial data banks. A phenomenal amount of information is available in these storehouses of learning, but sometimes the sheer magnitude of them makes it difficult to find what you're looking for. This chapter will show you what to look for and how to find it in the most expeditious manner.

The first time you research the literature for your article, it should be a cursory look, preliminary research, just to make sure that the subject you have chosen has not been overdone and to determine if enough literature exists to form the basis for a full-length article. This first look will give you enough information to write a query letter to interest an editor. After you receive your go-ahead from your editor, you can conduct your in-depth research to collect the information needed to write your full-length article.

The principles discussed in this chapter will help you in both searches. The major difference is that you will probably spend less than an hour on your preliminary research and perhaps a few hours on your in-depth research.

WHAT RESEARCHERS NEED TO KNOW

Any researcher needs to know two things:

1. Where to look.
2. How long to look.

This chapter, along with appendix A, shows you where to look. The "how long" is up to you.

When you're about to write an article and want to know what others have done in a similar vein, you need to research the literature. Knowing where to look and what to look for can save you much valuable time.

Each year, skilled librarians and abstracters spend hundreds of thousands of hours categorizing, cataloging, indexing, referencing, and abstracting the thousands of periodicals and books that are published that year. This chapter provides some paths to follow to help you locate the specific answers you need in the books and periodicals in this veritable mountain of information.

Books are the most important reference for research, but they are often out of date. Periodicals contain the most up-to-date information, but they generally do not cover subjects in adequate depth.

A LITERATURE SEARCH

The reasons for conducting a literature search are many and varied:

1. When you review established works, you're often learning from the best brains in the field.
2. Often you'll get new ideas, new slants on your article as you review other articles. Your creativity will be stimulated when you read how someone else attacked a similar problem.
3. In today's increasingly competitive technological world, you must be aware of, and hopefully leap-frog, what your rivals have done and have written about.
4. Another aphorism, "Two heads are better than one," applies. When you read how someone else handled a problem, you're reaping the results of two great minds working on the same problem: yours and the author's.
5. Billions of dollars have been invested in the projects that spawned those articles and books. Why not take advantage of this enormous investment?

A tremendous amount of information is available in those mountains of literature. And it's free. All it takes is a little time and effort, and some help in finding the proper direction for searching.

HOW TO GET STARTED

Your first step should be to that fount of knowledge, the library — either your company library or a university, college, or good-sized public library. (In chapter 9, on-line data banks, which are rapidly becoming a major source of information, will be covered.) If you have trouble locating the right library, the *American Library Directory* (ALD), Jacques Press/R. R. Bowker Co. (2 volumes) lists over 35,000 U.S. and Canadian libraries of all kinds: public, academic, company, association, etc. The directory is arranged alphabetically by state and province, and summarizes each library's holdings.

The library you select might not have on hand all of the documents you're seeking, but they generally have most of the basic reference books, abstracts, and indexes you'll need. If they don't have the book, periodical, or report on hand, they can probably order it for you through an interlibrary loan.

Start by checking the card catalog, which indexes all the holdings of that particular library. Many of the more modern libraries are in the process of converting to a computerized system where you can access the library's holdings from a computer terminal. If you have a computer at home, you can access some of these public libraries from your home or office.

No matter which system you use, the basics of finding information are still the same, so let's start with an old-fashioned manual card catalog, then we'll cover a typical electronic card catalog.

THE CARD CATALOG

Arranged alphabetically, manual card catalogs have an Author card, a Title card, and one or more Subject cards for each holding in that library. The card catalog indexes books, reports, pamphlets, periodicals, handbooks, transactions, etc.

An Author card is illustrated in FIG. 4-1. The articles *a, the,* and *an* are ignored in alphabetizing. Two basic methods are used in alphabetizing: word by word, and letter by letter.

The word by word system	The letter by letter system
New York	Newark
Newark	New York

Delivery Decimal System

Many libraries use the Dewey Decimal System, in which books are catalogued from 000 to 999, according to subject:

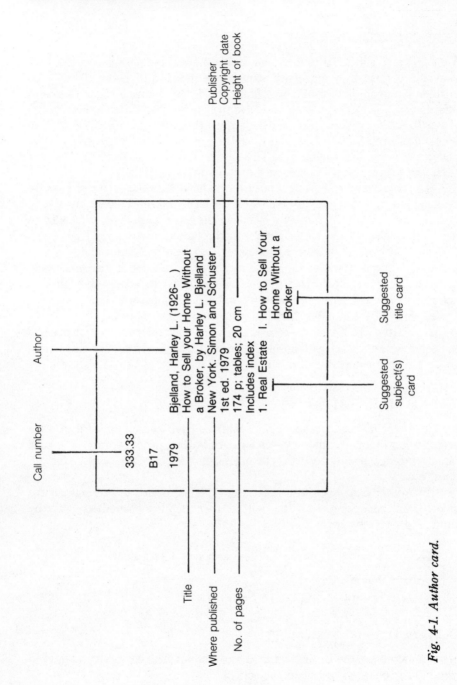

Fig. 4-1. Author card.

000   Generalities
100   Philosophy and related disciplines
200   Religion
300   Social Sciences
400   Language
500   Pure Sciences
600   Technology (Applied Sciences)
700   The Arts, Fine Arts, and Recreation
800   Literature (Belles-lettres)
900   General Geography and History

The Title and Subject cards have basically the same information, but in a different order. They are filed by title and subject, respectively.

Subject bibliographies of interest to the scientist and engineer are:

016.5   Science
016.6   Technology

Reference books are filed under:

030

Of primary interest to engineering and science professionals are:

500   Pure Sciences
600   Technology (Applied Sciences)

Each of the 500 through 600 codes is further broken down:

510   Mathematics
520   Astronomy & Allied sciences
530   Physics
540   Chemistry and Allied sciences
550   Sciences of the Earth & other worlds
560   Paleontology
570   Life sciences
580   Botanical sci.
590   Zoological sci.
610   Medical sciences medicine
620   Engineering & applied operations
630   Agriculture & related technology
640   Home economics & family living

650 Management & auxiliary services
660 Chemical & related technologies
670 Manufactures
680 Manufacture for specific purposes
690 Buildings

Each of these subjects is further broken down, for example:

621 Applied physics
622 Mining and related operations
623 Military and nautical engineering
624 Civil engineering
625 Railroads, roads, highways
626 Unassigned
627 Hydraulic engineering
628 Sanitary municipal engineering
629 Other branches of engineering

These numbers are further subdivided, ad infinitum. As an example:

621   Applied physics
621.3   Electromagnetic and related branches of engineering
621.36   Applied optics
621.361   Applied spectrometry
621.3612   Infrared

The Dewey Decimal System is used in most libraries because it can be easily adapted to the needs of a small book collection.

Library of Congress System

The Library of Congress system is used in most large libraries. Also, many science and engineering libraries are converting to it. The Library of Congress system allows for greater subdivisions, without making lengthy class numbers such as the 621.3889 in the Dewey Decimal. The classifications of primary interest to science and engineering are:

Q   Science
R   Medicine
S   Agriculture — Plant and animal industry
T   Technology
U   Military science
V   Naval science
Z   Bibliography and library science

The Library of Congress system is also further subdivided, for example:

T   Technology
TA   Engineering. Civil engineering
TC   Hydraulic engineering
TD   Environmental technology. Sanitary engineering
TE   Highway Engineering. Roads and pavements
TF   Railroad engineering and operation, including street railways and subways
TG   Bridge engineering
TH   Building construction
TJ   Mechanical engineering
TK   Electrical engineering. Electronics. Nuclear engineering
TL   Motor vehicles
TN   Mining engineering
TP   Chemical technology
TR   Photography
TS   Manufactures
TT   Handicrafts. Arts and crafts.
TX   Home economics

How to Use the Card Catalog

You can search for your topic by any combination of subject, author, and title. When you locate something of interest, be sure to write down the *full call number,* the *title,* and the *author* of the book or periodical in your index card notes. This not only helps you locate the specific book on the shelves, it also gives you a general location to browse through on the nearby shelves for related books on the same subject. Also, if the book you want has been checked out, the librarian will need this information to reserve it for you.

ELECTRONIC CARD CATALOGS

Libraries are gradually shifting over to a computerized card catalog. Many are converting only their more recent acquisitions and are gradually adding their older acquisitions to their computer database. So the total conversion may take a number of years to be completed, but eventually the manual card catalog will be mostly a museum piece.

Computerization of the library's holdings offers a number of features not practical in a manual card system. You can usually find what you're looking for in a shorter time, and you can search by author, title, subject, and call number. Also, you can also search using wild cards when you know only part of a title, name, or subject. In addition, the computerized system can tell you if the book is checked out, and if so, when it is due back—excellent features also not practical in a manual system.

As an example of the use of the computerized filing system, this is the procedure I used to call the Eugene, Oregon, library to obtain their on-line card catalog. My inputs are in boldface.

After my modem dialed the library's number, I received a message:

1200 connect

that told me that the connection had been made from my home to the library and that I would be conversing with the library at a 1200 baud rate. I hit my Enter key and the following series of conversations then took place:

USER NAME: EPLTEL
PASSWORD: (I TYPED IN THE PASSWORD)
         WELCOME TO THE EUGENE PUBLIC LIBRARY VAX 11/750
YOU MAY FIND MATERIAL IN FIVE DIFFERENT WAYS:
   BY TITLE   EXAMPLE   TI-HAWAII
   BY AUTHOR AND TITLE   EXAMPLE   AT=MICHENER/HAWAII
   BY SUBJECT   EXAMPLE   SU=HAWAII
   BY CALL NUMBER   EXAMPLE   CD=919.69 C226
   PLEASE TYPE IN YOUR INQUIRY AND PRESS THE RETURN KEY.

   I typed in **SU=TECHNICAL WRITING**

THERE ARE 12 TITLES IN THE CATALOG FOR SUBJECT TECHNICAL WRITING. BELOW ARE LISTED THE FIRST 10 (To conserve space I've listed only three.)

1.   Technical Writing (12)
2.   See also English language—Technical English
3.   See also Science news (2)
4.   Etc.

To continue, SELECT ONE of the CHOICES BELOW.

   I then input **1** as my choice and hit my **ENTER** key. I then received the following message:

1. 657.1 884   Crouch, William A Guide to Technical
2. 808.066 D661   Dodds, Robert H. Writing for technical Wri
3. 808.066658 H367h   Helgeson, Donald Handbook for Writing Pro

Press 1 to see more TITLES
Press 2 to display PREVIOUS SCREEN
Press 3 to display FULL BIBLIOGRAPHIC data for one of the above titles.
Press 4 to display LOCATION and AVAILABILITY of one of the above titles.
Press 5 to TERMINATE this inquiry.

I then input a 3 and a 1 and received the following message:

808.066658	Helgeson, Donald V.
H367h	Handbook for writing technical proposals that win contracts/
	Donald V. Helgeson..
	Englewood Cliffs, N.J. : Prentice-Hall, c1985
	xxii, 218 p. ; 25 cm.

Includes Index

Next, I wanted to check if the book was available and hit 2 to display the previous screen then I hit the 4 key and the following message appeared:

LIBRARY	LOCATION	VOLUME	AVAILABILITY
EUGENE	SHELVES		ON LOAN DUE BACK
			06 MAY

Seeing that the book was checked out and that I had saved a trip to the library, I then logged off.

Whichever system is available to you, browse through either or both the computer or the manual card catalog for your first effort. It's a good place to start.

But the card catalog is only a beginning. Next broaden your search into other areas of interest for additional information.

OTHER SOURCES

To expand your search, consider all possible documents, even those not in your library. These documents may be classified as shown in FIG. 4-2.

General References

For a general, broad picture of a topic, to help orient yourself at the start of a literature search, general reference books are often a great help. Usually these books do not have up-to-date information, but they do give you good background material. They also list additional references to check. General reference books are listed in appendix A.

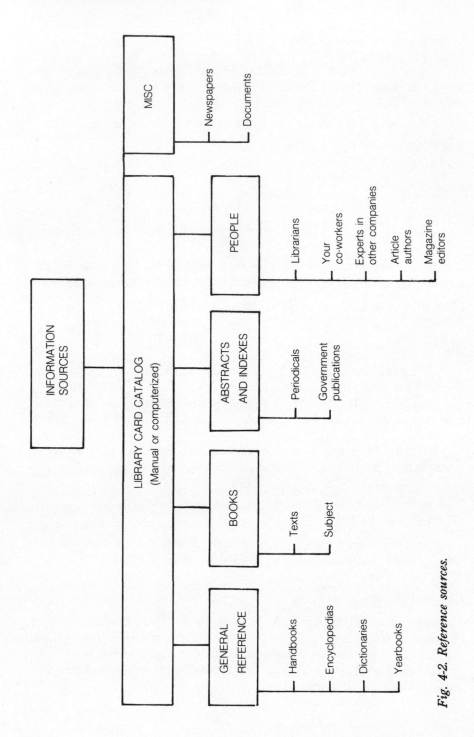

Fig. 4-2. Reference sources.

BIBLIOGRAPHIC REFERENCES

To list bibliographical references in your article, use the following format:

Books
Lastname, firstName. *title of book.* New York: Simon and Schuster, 1979.
Periodicals
Lastname, firstname. article title. *"periodical title."* 29 (June 1999): 1–4. (29
is volume number, and 1–4 are the pages the article is included on.)

Books on Reference Sources

If you have no idea where to start looking for references, you can begin with
the following sources. They are listed alphabetically by title, followed by the
publisher and years published.

A World Bibliography of Bibliographies, (1965–) Besterman. 5v.
Includes 117,000 items, grouped under 16,000 headings and subheadings. A
classified bibliography of separately published bibliographies of books, manu-
scripts, and patent abridgements. International in scope.

Bibliographic Index (1938–) Wilson. A cumulative bibliography of
bibliographies. An alphabetical subject arrangement of separately published
bibliographies, and of bibliographies included in books and periodicals. About
2,600 periodicals are examined regularly.

A Brief Guide to Sources of Scientific and Technical Information,
(1980) Information Resources Press. A concise, selective guide to sources of
information for the engineer and scientist. Lists principal science libraries in
the United States.

Government Reference Books, (1970–) Libraries Unlimited. Annotated
list of bibliographies, directories, dictionaries, statistical works, handbooks,
almanacs, and similar reference sources published by the U.S. government.

Guide to Reference Books, Eugene P. Sheehy. Chicago; (1986–) American Library Association. Subject, title, and author index of sources in all fields.
Special section on pure and applied sciences. An excellent, comprehensive
guide. The best!

Guide to Reference Material-Vol. 1: Science and Technology A. J. Wal-
ford (ed.); (1973–77) The Library Association, London 3v. Excellent, comprehensive lists of reference material in science and technology. Worldwide listing,
Emphasis on English-language materials.

How to Find Out About Engineering, (1972) Pergammon Press. 271 p.
Guide to various sources in engineering and its many branches. Covers bibliographies, encyclopedias, dictionaries, use of libraries, and standard reference
sources. Name and subject indices.

New York Times Guide to Reference Materials, (1985) Popular Library.

Shows where to look first when all you have to work with is a name. Covers a wide range of references; also has excellent sections on how to find information.

Science and Engineering Literature: A Guide to Reference Sources, (1980) Libraries Unlimited. General guide to sources of science and information fields. Over 1,200 reference sources covering basic reference books in general science, mathematics, physics, chemistry, computers, astronomy, geology, biology, engineering, and medicine.

HOW TO TAKE NOTES

Now that you know where to locate your information, you should adopt a good note-taking procedure. The method I use is described here.

The best medium on which to record your research notes is index cards. One side is normally lined; the other side is blank. They're stiff and easy to manipulate later when you organize them. They're available in 3-X-5, 4-X-6, and 5-X-8-inch sizes. Choose a size that's big enough to contain your notes and stick with that size. The 3-X-5 size is most commonly used.

Some general guidelines to observe when you take notes are:

1. Choose either the ruled or unruled side of the cards for writing your notes. (I prefer using the unruled side for my notes since I occasionally incorporate a simple figure or illustration along with my notes.) Whichever side you choose, be consistent — write your notes on only one side of the card.
2. Put only one idea, one topic, on a card. That way they'll be easier to organize later into topics and subtopics using the precedent sort. An *idea* or *topic* can be defined as any small amount of information that will not have to be broken up so that the parts can be placed at separate points in your outline. At the top of each card (see FIG. 4-3) put

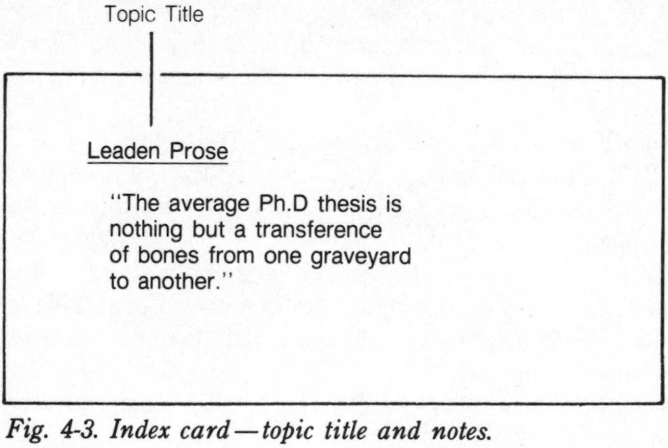

Topic Title

Leaden Prose

"The average Ph.D thesis is nothing but a transference of bones from one graveyard to another."

Fig. 4-3. Index card — topic title and notes.

a title, or topic, or subject, or heading or whatever you'll need for writing your notes and reviewing them later.

3. Use the opposite side of the card (see FIG. 4-4) to record bibliographic information. If the note is your own idea, record this fact on the bibliographic side. If you don't, when you've accumulated a lot of notes, you might later have difficulty remembering which ideas came from you and which came from the references.

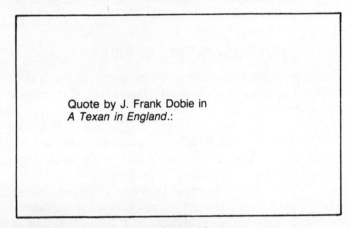

Quote by J. Frank Dobie in
A Texan in England.:

Fig. 4-4. Index card—bibliographic data.

4. Take many more notes than you need. Later you can prune them down to the essentials.
5. Be careful of using too much shorthand; you might forget what you meant by it. Record enough information to save yourself a trip back to the library.
6. Each card should have the following information as a minimum:
   a. The book or periodical reference
   b. Topic or subtopic title
   c. Your notes

7. Read through your references carefully to make sure you understand them, then digest and rewrite what you've read in your own words. Write down key ideas, words, and phrases.

When you do your research, you'll be taking three kinds of notes:

- Quotations
- Paraphrases
- Personal comments

Copy verbatim only if your reference has said something unique. You'll find that for 90 percent of the time you take notes, you will be paraphrasing the information. When it's in your own words, it suits your purposes better than the quotation, and you can usually shorten it.

If you must use a quotation, such as from an expert or well-known person, put quote marks around it, even if you paraphrase it later. Copy it exactly, including punctuation, spelling, capitalization, and paragraphing. If you omit part of the quotation that does not apply to your topic, indicate the omitted part with ellipses (. . .).

HOW TO REVIEW BOOKS QUICKLY

You're not going to have time to completely review all of the books you'd like to research, so there are some shortcuts that will give you enough information for your article:

1. Skim the book. Read the chapter titles and the bold headings, and look at the illustrations. They'll give you a quick summary of the book's contents.
2. Read the title page; the book may also have a subtitle. Note the author and his or her qualifications. Check the publication date to see how recently it's been published or if it's only a reprint of an earlier edition. Note the publisher. The book is more likely to be an authoritative reference if the publisher specializes in that type of book.
3. Read the foreword, preface, and introduction; they often give the purpose and intent of the book.
4. Review the table of contents; it outlines the book's contents.
5. Read the opening paragraphs of each chapter.
6. Check the index for topics of interest and see how many pages are devoted to these topics. If you find some topics that pique you, look them up, skim them, and make any notes of topics that interest you.
7. Check the bibliography. It will reveal the author's sources and whether he or she is using up-to-date information.

EXERCISES

1. Look up the periodicals of your choice in either Gales Research Cos. *Book Review Index* or *Ulrich's International Periodical Directory.* Record any pertinent information.

2. Review at least six books or articles in periodicals that cover a similar topic to the one you have chosen.

3. For these references, critique the published material. Why is your proposed article going to be better than those already published? Is yours more recent? Does yours provide a better coverage? Does it cover items not included in the other articles?

5

Organize Your
Material Properly

*Order and simplification are the first steps
toward the mastery of a subject . . .*

Thomas Mann

In a major city, a four-story, red-brick, office building had to be moved to a vacant lot twelve blocks away to make way for a freeway. Three movers were invited to a conference held by the building's owners. They presented their credentials to be awarded the job.

Ace Movers, the biggest of the three, won first chance. Ace sent their engineers and mechanics to the location and they immediately tried to lift the building with jacks and move it to its new site. Although they labored long and hard, the jacks were not sturdy enough, and broke each time they were used to try to elevate the four-story structure. So this method had to be abandoned.

Modern Movers, the next largest in size and the second choice, decided that thinking hard was better than brute force. They hired a group of college professors and computer consultants who were full of ideas and good intentions. They concentrated deeply on ways to move the building. They computed the forces needed to move the structure using large blimps to lift it and carry it away. But it would have taken 10,000 giant blimps and wire harnesses, which would cost too much. As their computers spewed out calculation after calculation, the learned men also considered helicopters and dozens of other methods to lift it up all at once. But their calculating, thinking, and talking exhausted them, and they finally theorized that it was absolutely impossible to move the building.

SIMPLIFY

Walter Bell, of Bell Movers, the tiniest of the moving companies, had the next chance. Walter told his engineers to "Simplify." No one really knew what he meant by that, but he called his few employees together and accompanied them to the building. They walked around it, measured it, calculated, and came up with a plan. Their detailed plan was to start at the very top, carry one brick at a time to the new location, and reassemble the building according to the detailed sketches they had made. It was a very simple solution and it took many trips to move the building, but before long the building glistened in its new location in the middle of the formerly vacant lot. Bell and all of his employees earned special bonuses for solving the problem so effectively.

Writing an article, or a story, or a book is a similar project. It's too big a job to jump into without planning. It's much too practical a job to only think about, much less to try to do by good intentions alone. Yet, by reducing it to a detailed and well-thought-out plan, a series of small steps, and doing each of these steps, one at a time, even a small person can do a big job.

That is precisely what a good outline, a logical ordering of your topics and ideas, accomplishes for you. Organization reduces a huge, seemingly insurmountable project into a series of small, interrelated steps, each of which can be done one at a time and be joined together, until the total writing project is completed.

BRAINSTORMING

By this time you should have accumulated a number of notes pertaining to your topic. Now that you have built up your confidence by amassing a lot of background information, it's time to proceed to the next vital step: brainstorming.

Brainstorming is a process where you think about your topic uncritically, without giving any thought to organization or judging an idea's worth. You open the floodgates of your mind and write down every idea you have about your subject, in whatever random order they occur to you. Let the ideas gush out in any form, in any order. Don't try to evaluate your material at this point. Creative thinking requires a positive attitude, so write down all of your ideas, knowing for certain that some of them will be good.

Judging too early is the big enemy of brainstorming and must be avoided or it might cause you to cut off the flow of ideas or disregard promising ideas. You can produce ten times as many ideas as you could if you paused to judge each one when it occurs. Defer your judgment or you might evaluate and discard ideas that do not seem good, but which, through association with your ideas, could give birth to new ideas.

Evaluate and organize your ideas later. To brainstorm, you must let your brain roam free, unfettered, unbiased, unorganized, uncensored.

To start your brainstorming session, get the set of note cards you've accumulated from your library research and procure an additional supply of blank note cards. Start reading through your note cards and let your mind do a little creative wandering as you read them. This activates your mental muscles, setting the marvelous process of idea association to work.

Each time you read one of your note cards, one or two more ideas or thoughts will pop into your mind. Jot them down on separate blank cards, one idea or thought per card. A huge quantity of ideas is needed because quantity helps breed quality. Build new ideas and modify the ideas you've already written down. Write down all your thoughts, no matter how wild and impractical they seem. You are writing them down for only yourself to see, so don't be embarrassed by what you write.

Asking yourself questions is often an excellent way to get started because questions are the creative acts of the mind. Read through these questions slowly and see which you can apply to your topics.

Ask yourself, "Can I . . .

- Use a different shape, a different form?
- Make it faster, less expensive, smaller, lighter?
- Add new features to make it more versatile?
- Combine it with something else?
- Reverse it or change the sequence?
- Find new applications for it?
- Use different ingredients to obtain new properties?
- Adapt ideas from a related product or process?
- Split it up into smaller parts?
- Use a different process?
- Merge old ideas?

Brainstorming is not a five-minute task. It might take an hour or two, perhaps longer if you have a complex subject to discuss. But this is prime time. It gives you the basis, the framework for your article — the points, the unique ideas, the thoughts you are going to cover. This is truly the most creative part of your writing; it is the task that can make your article an outstanding one.

Go through your old and new cards a few more times. More and more ideas will occur. Write them down, uncritically, and you'll be amazed at how the stack of cards has grown into a full deck.

When you feel your creative juices have finally run their course, set all of your notes aside and do something else for a few hours. Take a walk, shoot some baskets, hit a few golf balls, do something physical to rest your mind and let your subconscious go to work. If you can spare the time, sleep on it. Next you're going to be ready to form the skeleton of your article, to organize it, and to make order out of chaos.

ORGANIZATION

The most important single key to effective written communication is good organization! A document properly researched and organized is half written.

In a survey of technical writers—professionals who earn their living from writing—only 5 percent said they used no outline for their writing. Most likely some of the inept technical writing you have encountered was written by this 5 percent.

Most writers agree that one of the most difficult tasks is organizing their work. In a recent survey 1000 engineers were asked what bothered them the most about writing. Leading the list of dozens of complaints at 28 percent was organizing and outlining. Some people equate creating an outline to having a dentist perform a root canal. But this book promises the novocaine to take the pain out of outlining, and to make it into an enjoyable, logical, and profitable enterprise.

It's difficult to overemphasize the importance of a well-constructed, comprehensive outline. The more time you spend on creating and revising your outline, the easier your article will be to write. While your ideas are still in outline form, you can review, evaluate, modify, correct, rearrange, even totally reconstruct them much more readily than a written draft of an article. You can easily spot missing or inconsistent topics in an outline and correct them. And you can spot and eliminate topics that aren't necessary. An outline gives you a track to run on. It's a beacon that lights the way when the going gets rough.

Once you start putting words on paper, it is extremely difficult to totally rearrange your ideas and still maintain continuity. A completed rough draft almost sets the article's contents and order in concrete, requiring a jackhammer effort to change it. Detailed outlines encourage forethoughts, rather than afterthoughts.

This chapter takes that very difficult, even onerous task of organizing and shows you how to organize your thoughts according to sequence, importance, problem/solution, or whatever order you elect to use. Introduced in this chapter is the precedent sort (p-sort), a revolutionary, new technique that performs this vital task of organizing, making it almost formulalike in its simplicity, precision, and ease of use.

You'll find that once your ideas are properly organized into an outline, your words, sentences, paragraphs, and pages fall into place effortlessly. When you write, it's essential to know what you're going to say next. The precedent sort, or p-sort, gives you this direction.

The P-Sort

A comprehensive outline makes your article easier to write because once you know that you've covered all of the important points you need not worry about being distracted by thinking about your other topics. You can concentrate your efforts totally on the topic you're writing about.

The p-sort shows you how to establish this direction. It works for articles, for chapters in a novel or a nonfiction book, in technical writing, in a memo, in speeches, in all types of writing.

If someone hands you a poorly organized report, it's as if you received a dozen beakers of chemicals that comprise one of your company's products, with no directions on how to process or mix them. All the chemicals (facts) are there, but they are of no use unless they are processed (organized) properly.

That's what the p-sort does for you. It provides the directions to assemble the vital parts of your articles in the proper order; to form a unified useful product that other people can understand.

Basic Writing Steps

Two basic steps in writing, whether it's for an article, a letter, a proposal, a report, whatever, are:

1. Research and organize your ideas
2. Write and revise

Many people skip the vital first step, dive right in, and start writing. Invariably they end up thrashing about, drowning in a rambling, circumlocutious document that is difficult to write and even more difficult to read and comprehend. All of the facts might be there, but unless they are arranged in a logical order, they can't make their way from the published paper into the reader's mind.

Our brains simply are not geared to think in a logical, deductive, chronological order when we're creating something. Our minds hop about, generating random ideas loosely connected by some inexplicable, associative, magical mental process. Each individual thought might be a brilliant idea by itself, but unless they're strung together into some logical, orderly pattern, the result is a meaningless jumble to anyone else.

To demonstrate this phenomenon, try this experiment. Sit down and write everything that comes into your mind for about five minutes. Then read it back and see how your mind skipped wildly about and how little sense your unorganized thoughts would make to some stranger trying to read and understand what you just wrote down.

OUTLINES

Proper organization of your material before you write:

- Makes writing easier.
- Lets you focus all of your efforts on only one topic at a time.
- Helps you include all of the points you want to make in a logical sequence.
- Enables you to start at any point in your article since you know what is going to be included in the entire document.

A good outline is a strong skeleton that you can later flesh out with words to create a well-formed, complete body. If the skeleton is not assembled properly, the body will not operate smoothly, but in jerky, disconnected movements.

Most outlining techniques you have used have given you a lot of confusing theory, replete with Roman numerals, but with very little practical help on this important endeavor. You've probably read about, and have suffered through, the Inverted Pyramid, the Suspense Formula, the way to indent and count by Roman numerals, ad nauseam, but no technique has shown you a simple way to organize and join your topics together.

Over the past twenty-five years I have evaluated, developed, and tested numerous methods of organizing, and have adapted and modified the Bubble Sort method used in computer programming to organize material for writing. I've named this method the Precedent Sort, or p-sort. It's a systematic, logical way to organize a large group of randomly generated subjects into an orderly whole. It shows how to gather the scattered bones and put together a properly constructed skeleton.

Topic vs. Sentence Outlines

Before the precedent sort is covered, it's important to discuss the two basic categories of outline forms that are used:

- Topic outline
- Sentence Outline

A *topic outline* lists the topic to be covered in each specific part of an article. This type of outline can be used when the specific meaning of each topic is clearly understood. For example, a topic outline would use the word *operation.*

A *sentence outline* provides a more thorough definition of what the topic covers. For example, instead of the word *operation,* a sentence outline would form a complete thought such as, "How does it operate in a severe heat environment?" A sentence outline clarifies the topic and makes the final writing easier and faster. Often the sentence outline can form the logical topic sentences for paragraphs, or the key statements for entire strings of paragraphs.

Both topic and sentence forms can be combined effectively in developing an outline. Use whichever best suits your purposes.

Ordering Sequences

Before you begin to organize your material, you need to decide what type of ordering sequence you will use. Outlines can be sequenced in a number of possible ways:

- Chronological
- Spatial
- Increasing detail
- Major divisions
- Inductive/Deductive
- Known to unknown (analogy)
- Increasing or decreasing importance
- Cause/Effect
- Comparison/Contrast (Advantages/Disadvantages)
- Literary or suspense

You'll be using one or a combination of these sequences in writing your article.

A *chronologically ordered* outline reports a series of events by arranging the topics in the sequence in which they occurred. You write about the event or the topic that occurred first, then second, etc., until you have reached the last event or topic. This type of outline is suitable for trip reports, processes and operating procedures, progress reports, lab reports, analyses, etc. (Chronological ordering was used to arrange most of the chapters of this book in the order you should normally follow in writing an article.) Chronological ordering, however, does not provide proper emphasis because the most important parts of an article might be buried in the middle. Also, there is a tendency to include unnecessary information to make sure the article covers everything.

Spatial ordering organizes topics in geographical, or physical order. It may be used in describing a plant, sales and marketing reports by geographical areas, or a company organization by divisions and departments. This type of outline is also relatively easy to write, however, it also does not place proper emphasis on the most important topics. In addition, information of dubious value might have to be included for completeness.

An *order of decreasing detail* is used in journalism. Newspaper articles begin with the general information everyone wants to know. Subsequent paragraphs add more detail until the writer runs out of facts or the editor runs out of patience and limits the space allocated to the article. A news story has to be written in this manner so it can be chopped off after any paragraph to be squeezed into the newspaper space with the article still complete and the gist of the story still included. This type of outline can be used to present information of value to a broad technical audience of readers, with varying levels of interest and backgrounds, such as in engineering or market research reports with a wide circulation.

The *major divisions order* is used for a topic that easily divides itself up into natural and obvious parts. For example, a proposal or lab report is often divided into the specific major steps dictated by the standards of the company involved. A description of an organization naturally follows this order since it is based on a company organization chart, describing the functions of the president down to the lower levels, and the various divisions and departments of a company.

The purpose of the *inductive/deductive order* is to convince the reader that the conclusions in the report can be arrived at by logical induction or deduction. The inductive order begins with specific observations and leads to a general conclusion. You present evidence that you believe will support your conclusion. For example, you begin by saying that ABC Company's chemical products are inferior, then add that their petroleum products are inferior, then continue to add more negative evidence, and conclude that ABC should be banned from all manufacturing. Specific instances of the inferiority of some of the company's products lead to a general ban of all the company's products.

The deductive order is the opposite and proceeds from general observations to a specific conclusion. You state a general proposition or viewpoint, then present evidence to support it. For example, you might start out by stating that all foreign automobiles are unreliable, then proceed to point out examples where this unreliability has been demonstrated so that the overwhelming weight of the evidence you present adds up to confirm your proposition.

When the material involves controversy, the inductive, or facts-to-conclusion, method is more convincing than the deductive. If you have a difficult selling job, use the inductive order. Induction leads the reader gently to your conclusion. The evidence is presented to gradually lead to a concluding statement, and can be accepted at face value by even a doubting reader. It makes the writer look unbiased, although you really aren't. You simply don't present evidence to counter your proposition and conclusion.

On the other hand, the deductive order risks alienating the reader who, resisting the initial statement, may view even the strongest evidence in a negative light. This order helps readers get to the point quickly, however, and saves time for top managers by putting the recommendations up front. With the recommendations incorporated at the beginning of the document, the rest of the document is used to justify the conclusions.

The known to unknown order is used in documents such as technical manuals when explaining new equipment or a theory of operation. The writer starts with facts the reader understands, then bridges to another fact that can be deduced easily from the first fact, and continues to link facts until the reader is informed of, and understands, the unknown. Often an analogy will help with this method. This method is often called the *professorial,* or *teaching,* method.

Items or the topic can also be arranged in *order of increasing or decreasing importance,* depending on the desired effect. An order of increasing importance, typical of mystery stories and scientific reports, presents the least important information first, then builds to a conclusion by presenting the most important information — the results — last. If done properly, this method can capture and keep a reader's interest all the way to the end. But it is also difficult to use because few topics lend themselves to an order of increasing emphasis.

An order of decreasing importance is typical of reports that start with a long abstract, or summary of conclusions and recommendations. This order puts the most important facts or data first, to give the reader a quick review of points, then continues down the importance ladder to the bottom rung, the least important. This order is preferred by managers who want to read the results first. However, because of the decreasing level of interest, many readers will drop off on the way to the end of the article.

Cause/effect is used to explain what forces or events produced particular results. And, if the effect is not a desired one, such as in a problem/solution, the conclusion is a recommendation of what should be done to prevent a recurrence of the wrong result. It is also used for proposals and feasibility studies.

Comparison/contrast compares or contrasts two or more items, such as leasing costs versus purchasing costs or the performance of one company's products against another's, using an advantages/disadvantages comparison. This type of order is repetitious, however, unless comparisons and contrasts are done efficiently through the use of tables and graphs, rather than text.

The *literary* or *suspense* formula reveals a little information at a time and builds up a suspense to a hopefully unexpected and entertaining climax at the end of the article. Humor in nonfiction uses this order effectively.

WHICH ONE SHOULD YOU USE?

Depending on the purpose of your article, you can organize your material in one or more specific ways:

- *To inform, document, or entertain*
  - Chronological
  - Spatial
  - Major Divisions
  - Increasing Detail
  - Increasing or Decreasing Importance
  - Literary or Suspense

- *To convince or to get action*
  - ○ Inductive or Deductive
  - ○ Problem/Solution

- *To explain or to get understanding*
  - ○ Known to Unknown
  - ○ Comparison/Contrast

THE PRECEDENT SORT

Art is nothing without form.
Gustave Flaubert

Without some methodology, it's difficult to evaluate twenty to fifty or more randomly listed topics simultaneously and decide, by trial and error, the order in which they should be arranged. It's like the kid in the candy store, who said, "It's easy to decide if I want to buy either chocolate drops or peanut brittle, but when they add jelly beans, salt water taffy, and caramels, I get all confused."

The precedent sort (p-sort) puts order into this task. It breaks down a complex organizing task into a series of one-on-one decisions and shows you how to compare topics or ideas, one pair at a time, and select the topic or idea that has precedence; that is, the one which should come first. You continue until all comparisons are made and *violà*, your organization is complete.

Now that you have ten to fifty topics listed on your index cards, you can see the difficulty in trying to sort them into some meaningful order, unless some system is available to assist you.

Principle of Fewness

The Principle of Fewness states: the fewer items that are presented as a group, the easier they are to evaluate and understand. So the task is to reduce this group to as small a group as possible, one that can be comprehended and dealt with easily. Enter the p-sort. The p-sort optimizes the Principle of Fewness by reducing the number to be compared to a minimum: two. You compare only two items at a time, evaluating one against the other to determine which you should cover first in your article.

An Example The p-sort is best illustrated by a simple example. This example will use numbers to illustrate the principle. Next an example of writing a technical article will illustrate the application of the p-sort to a practical writing problem. I recommend that you take a set of 3-X-5 cards, write the listed numbers on them, and stack them in the order shown here. That way you'll obtain some good experience with the technique.

Your set of five 3-X-5 cards should each have a number on it and be arranged in the following order:

Card A   1941
Card B   187
Card C   1001
Card D   456
Card E   26

Note that it is not necessary for you to identify the individual index cards with letters as used in this example. I used the letters only so I can refer to them in the text when explaining the p-sort process.

Your task is to arrange these numbers in ascending order. You probably could do this quicker by sorting the index cards by hand, but this example has been made short and simple to illustrate the basic principles of the p-sort.

First, place all of the cards faceup in a stack and pick up the first card, Card A. With your other hand, pick up Card B and compare the two, as shown in FIG. 5-1.

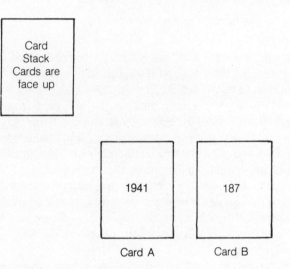

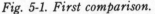

Fig. 5-1. First comparison.

Card B has a smaller number than Card A, so place Card A facedown in a new pile we'll call the Unordered pile and hold Card B. You are going to work your way through the entire stack a number of times. If the next card you pick up is lower in value than the one you're holding, you'll place the higher numbered card facedown on the Unordered pile and continue to hold the lower-numbered card.

To continue the first round, pick up the next card from the top of the stack and compare Card B with Card C, as shown in FIG. 5-2. Since 187 is smaller than 1001, continue holding Card B. Place Card C facedown in the Unordered pile on top of Card A.

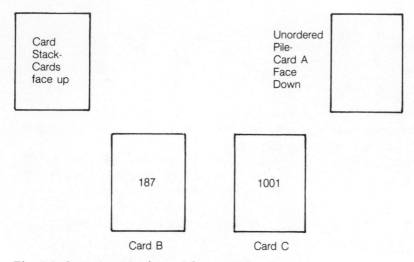

Fig. 5-2. Second comparison of first round.

Pick up the next card from the stack and compare Card B with Card D. Again, 187 is smaller than 456, so place Card D facedown on the Unordered pile and continue to hold Card B.

To complete the first round, pick up the last card from the stack and compare Card B with Card E. Since 26 is smaller than 187, place Card B facedown on the Unordered pile. Finally place Card E facedown in a new pile we'll call the Ordered pile. The smallest number (26) has now "bubbled up" to the top and is the lowest number of all the cards.

This procedure is akin to a computer programming sorting routine called the *bubble sort*. As you go through the remaining unsorted cards, successively smaller numbers will bubble up to the top and be placed facedown on top of Card E in the ordered pile. One number will join the ordered pile after each round of comparisons.

To start the second round, turn the unordered pile faceup, pick up the first card, Card A, and compare it with Card C, as shown in FIG. 5-3. Since 1001 is smaller than 1941, place Card A facedown on a new Unordered pile and pick up Card C.

Compare Card C with Card D, as shown in FIG. 5-4. Since 456 is smaller than 1001, place Card C facedown on the Unordered pile and hold Card D. Continue the process until you have compared all the cards. At the end of the second round, you should be holding Card B. Place Card B facedown on top of Card E on the Ordered pile. So far two cards, Card E-26 and Card B-187, have bubbled up to the top.

To start the next round, turn the Unordered pile faceup again and repeat the procedure until you have sorted all the cards. For five cards ($n=5$), this should require $n-1$, or $5-1 = 4$ rounds of comparison.

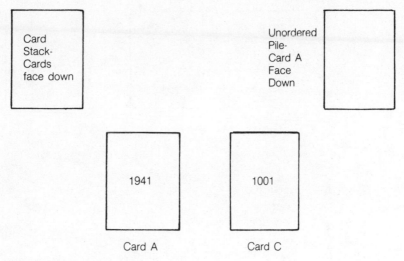

Fig. 5-3. Start of second round.

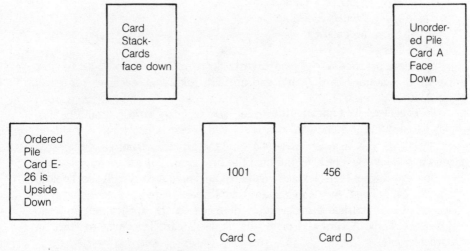

Fig. 5-4. Second comparison, Round 2.

The final order is shown, left to right, in FIG. 5-5.

A More Practical Example For the next example, an inductive order of organizing will be used, arranging the topics for an article in a periodical to convince the reader to buy your product. Assume you are going to write an article for a periodical about a new device you have just developed for your company. The topics you want to cover include the following. These were written down in a random order and will be organized using the p-sort.

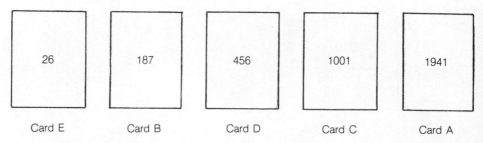

| 26 | 187 | 456 | 1001 | 1941 |

| Card E | Card B | Card D | Card C | Card A |

Fig. 5-5. Final sorted order.

Topic A	Applications
Topic B	Price
Topic C	How does it work?
Topic D	Reliability
Topic E	Future developments of product
Topic F	Self test
Topic G	Background of product
Topic H	Field test results
Topic I	Models available
Topic J	Modularity

For more practice on the p-sort, write down each topic from FIG. 5-6 on a separate 3-X-5 index card. You'll end up with ten cards, each with a single topic written on each.

To organize the topics in the order you're going to use, compare the topics listed in FIG. 5-5 one pair at a time, and decide which topic you should logically discuss first in your article. To start the procedure, put all of the cards faceup in a stack, then pick up and hold the first card, Topic A — Applications. Pick up Card B from the stack and compare the Topic A card with the Topic B card. Logically Topic A — Applications should be covered before Topic B — Price can be established because price depends on the application, so keep holding the Topic A card and place the Topic B card facedown to start an Unordered pile.

Pick up the next card on the stack and compare the Topic A card with the Topic C card. Logically, Topic C — How does it work? should come before Topic A — Applications can be understood, so place the Topic A card facedown on top of the B card on the Unordered pile and hold the Topic C Card.

To continue, pick the next card off the stack and compare the Topic C card with the Topic D card. Logically, Topic C — How does it work? should be known before Topic D — Reliability can be established, so place the Topic D card facedown on the Unordered pile and hold the Topic C card.

Continue this process, always placing the card bearing the topic that should come later facedown in the Unordered pile and holding the card with the topic that should be covered first. When you reach the end of the stack, you should have Topic G — Background of product in your hand. All the other cards will be facedown in the Unordered pile. So, at the end of this round, place the Topic G — Background of product card facedown in a new pile, the Ordered pile. Topic G has bubbled up to the top and has become the first subject that will be discussed in your article.

To start the second round of comparisons, turn the Unordered pile faceup and pick up the first card, Topic B — Price. Pick up the next card from the stack and compare Topic B with Topic A. You should conclude that Topic A must be discussed before Topic B can be established, so place the Topic B card facedown to start a new Unordered Pile and continue to hold Card A.

Next, compare Card A with Card D. You should conclude that Topic A should be covered before Topic D, since reliability depends on the application. So place the Topic D card facedown on the Unordered pile and continue to hold Card A.

Continue going through the stack. By the time you reach the end of the second round of comparisons, you should be holding the Topic C card. Place this card facedown on top of the Topic G card on the Ordered pile. Now you have two topics, Card G and Card C, that have bubbled up to the top to become the first two topics you will discuss in your article.

The first-to-last order I came up with after all of the topics had been compared and bubbled up to the top follows:

Topic G	Background of product
Topic C	How does it work?
Topic A	Applications
Topic J	Modularity
Topic F	Self test
Topic H	Field test results
Topic D	Reliability
Topic I	Models available
Topic B	Price
Topic E	Future developments of product

Note that although this is a "formula," you must still make subjective judgments in all the pairings, and your opinion of what should be covered first might differ from mine. So you see that this technique has room for individual creativity and that's precisely why it is so universal in its applications.

The basic principle of the p-sort is to put blinders on — to compare only two topics at a time, then concentrate and select the one topic you should cover first in your article. Thus the method of organization is broken down from the mind-boggling, complex problem of simultaneously trying to compare and organize twenty to fifty topics, to one of comparing only two at a time.

Spend all of the time you need in making the comparisons — minutes, hours, or longer if needed. If your organization doesn't come out the way you like it after going through all the cards, start over and repeat the process as many times as you need to come up with an order that satisfies you. You'll know you have the right order when you can go through all the cards and not need to change the order of any of them.

You can use the same technique to organize:

- Chapters in a book
- Topics within each chapter
- Events in short story
- Topics in a technical manual or report
- Events in a novel

Filtering Your Ideas

Now that you have all your cards in order, you're in a position to filter out unnecessary information. Up to now you have deliberately restrained or suppressed your critical judgment and evaluation of your ideas. Now it's time to be critical, to make judgments, to filter your notes.

To prepare to judge, find a quiet corner and go through your note cards, one by one. As you read through them, it will soon become obvious that some of the notes don't belong in your article. If, as you review them, you are suddenly jolted by a note that doesn't seem to belong, set that card aside and continue reviewing. Remember, you were encouraged to take more notes than you needed, so don't be afraid to filter the surplus ones out. Continue going through your note cards, setting aside any that don't belong.

Go through your note cards again and see if they read better this time. If you again have a jolt in continuity, perhaps you should put one or more of the cards you discarded back in your Ordered stack. If you feel something is missing, make a note of it on one of your blank cards and insert it where it's needed.

Repeat this process a few times until the notes form the pattern you want—until they're right, with no jolts in continuity, no missing steps.

Remember, organizing is not a static process; it's a dynamic one. So continue to make any changes you feel are necessary at any time. Your index cards are ideally suited for this procedure. You can move them around, add to or subtract from them, and totally reorder them until you are completely satisfied with your organization.

Outline Formats

Most people with whom I've discussed outlines have very unpleasant memories of the Roman-numeral outlines that were force-fed them in high school and college. So we're not going to dredge up those unpleasant memories. We're going to devise a simple outline procedure that you can adapt to your purposes. This technique uses no Roman numerals whatsoever.

By this time you should have your note cards arranged in the proper order, so it's time to place them all in a single row, glance through them to see how they look, then arrange them according to topics, subtopics, etc.

Line up all your cards in one or more vertical rows. Then, by simply moving some of them to the right and left, arrange them so that the subtopic cards are physically indented from the topics and that the sub-subtopics are physically indented further from the subtopics, etc. This will give a physical outline of how they all line up and are subordinated.

The best method is to study each card in order and decide if it has the same *weight,* or the same importance, as the topic above it. If the lower card has the same importance, don't move it. If it is a subject of less importance than the one above it, physically move it to the right. If a topic is of more importance than the one above it, physically move the topic card to the left. This physical method of indenting your topics gives you an excellent overall picture of your entire outline.

The system I use at this point, once I have my cards laid out in a physical outline, is to type these topic titles into my word processor, using the following coding and indentations. I copied only the topic titles from my cards.

ORGANIZE YOUR MATERIAL

Movers and Shakers (Anecdote)

Brainstorming

- Uncritical thinking
- Review note cards

- - Add new ideas as they occur
- - Modify existing ideas as required

- Not a five-minute task
- Continue until creative juices cease to flow

Organization

- Single, most important key to good writing
- Difficult task

- - 28% say most difficult of all writing tasks

- Time spent in organizing saves writing time

- - Less confusing
- - Spot items left out
- - Spot surplus items
- - Encourage forethoughts, rather than afterthoughts
- - Simplify writing

Organize Your Material Properly

So, for an article in which the topics and subtopics are limited, a simple code suffices. I underline a major topic, indent and use one dash for a subtopic, and indent further and use two dashes for a sub-subtopic. It's a simple but effective way to outline.

At this point in your article you can transfer the topic titles from the top of your note cards into your word processor or to a few sheets of paper, complete with the proper subordination. Don't copy all of the notes on the index card; copy just the topic title at the top of your card. Now you have what we'll call a *working outline* of your article completed.

Next?

The p-sort lets you order any number of topics (you can expand the list of topics up to perhaps fifty or more) in any desired sequence, which makes your writing flow logically and easily from one topic to the next. It's much easier to concentrate on and compare only two topics at a time and decide which comes first, rather than trying to evaluate fifty randomly arranged topics at once and trying to arrange them by trial and error.

You can use this valuable technique to organize the subjects in a memo, the topics in a long technical report, or the subjects in a proposal. It's a powerful technique. The more you use it, the easier it becomes.

Guidelines in Using the P-Sort

Some important guidelines to observe when using the p-sort:

1. Proceed logically from the top to the bottom of the Unordered pile. Don't jump around; your results will be confusing and the comparisons will be more difficult.
2. Don't decide that it's not important which topic comes first in a pairing. Make a judgment for each comparison, weighing all of the factors as best you can. If all else fails, just guess, but you must make a decision or the process won't come out right. If they're that close, it probably doesn't matter which comes first. Besides, if you change your mind on the order later, it's easy to modify.
3. Before starting to make your comparisons by pairs, make sure you understand precisely what each topic is and means. This is very important. If you're not sure, write out a statement or two, or an entire paragraph, on the card to accurately define each topic before you start your comparisons. There is no restriction on what you can write on a card, so define it so that you know exactly what it means.
4. To organize a huge list of items, say fifty or more, first sort the index cards into smaller groups of related items. Then take each group, one at a time, and order each group by itself using the p-sort. Finally, use a master p-sort to organize the order of the groups.

A Caution

Don't conclude that the p-sort is a mechanical, computerlike way of organizing topics because it isn't. You still must use your judgment, your creativity, just as you do when you use all formulas. The p-sort simply helps you break down a huge, perplexing problem into a series of simple, orderly, logical decisions, then helps you organize the results.

Also, the outline you have created here need not be cast in iron. If there are compelling reasons to change the order around, do so. It's a lot easier to work with and manipulate an outline after it's "almost organized," than when you're trying to organize fifty randomly listed topics.

OTHER APPLICATIONS OF THE P-SORT

You can use the p-sort for many other applications such as how to:

1. Schedule tasks, itemizing the order in which jobs must be accomplished to reach a goal in a logical, efficient manner.
2. Rank the topics in a sales brochure, listing the most important features first.
3. Arrange a large group of items in numerical, alphabetical, or chronological order.
4. Assign jobs. Use two precedent sorts for this application: one set to rank your employees in order of their competence, the second to rank the jobs to be done in their order of difficulty. Finally, assign your best employee to your hardest job, your second best to the second-most difficult task, etc.
5. List your own jobs to be done in your daily life, in a required time or spatial sequence, with the hardest jobs first, or in whatever order you choose.
6. Arrange ideas that must be sold or justified in order of difficulty, such as in a proposal where you should spend the most time discussing difficult problems that are to be solved.
7. Arrange a big decision into a series of little decisions that must be settled before the big decision must be made.
8. Judge contests, ranging from speeches (ranking the speakers) to a cake-baking competition, to beauty contests.

The next time you have a document, an article, a book, a report, or a proposal to organize, use the p-sort. You'll find it to be a capital idea.

EXERCISES

1. Why should you outline your articles before you write?

2. Get a set of 3-X-5 cards (or whatever size you prefer) and use the p-sort to alphabetize this list:

   Chemist
   Mechanical Engineer
   Architect
   Executive
   Scientist
   Writer
   Technician
   Civil Engineer
   Physician
   Programmer

3. How many passes did it take to organize the group in Question 2?

4. Organize the topics of your article using the p-sort.

6

Visuals

A picture shows me at a glance
what it takes dozens of pages
of a book to expound.

Turgenev

Good visuals can make or break an article. People understand and remember images and pictures, not abstract words. Visuals can clarify images too complex to be conveyed by language. Imagine the impossibility of fully describing a Van Gogh painting in words alone.

It's important to tentatively plan what visuals you're going to use *before* you write your text so that the visuals form an integral part of your article, rather than an appendage tacked on as an afterthought. Visuals must mesh with your text, like two gears that drive a machine. They must work in concert, each dependent on the other, to describe an object, a process, or a concept.

Because readers generally do not study visuals in detail, the text accompanying the visual should include enough information to help the reader understand the significance of the visual.

You can modify, add to, or delete visuals later when you write to a plan. It's much easier to modify a plan than to work with no plan at all.

WHY USE VISUALS?

Visuals—a term encompassing line drawings, graphs, photographs, pie charts, curves, and tables—serve many functions in an article:

1. They are often used to explain difficult concepts that are impossible to put into words. With text each word is read in series, one word after the other, and it takes many words to convey a thought or to create a complete "picture" in a reader's mind. With a visual, the whole picture is available at once, in parallel, and the reader can see how the various components of the picture relate to each other in size, in position, and in function.

2. Visuals provide eye relief, a needed rest for the eyes, by breaking up the solid, imposing text of an article with white space and a different media of presenting information. This pause helps readers better absorb the content of your article.

3. Technical people are accustomed to visuals and spend much of their professional life working with engineering drawings, curves, graphs, tabular data, etc.

4. Technical subjects generally deal with concrete things, which can usually be better explained with visuals.

5. Visuals increase reader retention of your article since visuals stick in your reader's mind much longer than words. It's been proven that we generally remember only 10 percent of what we read. However, if a visual accompanies the text, we remember 30 percent of a written message. That's a 300 percent increase!

6. Everyone who sees well-drawn visuals sees the same "picture." But words are abstractions and mean different things to different people. If you use the simple word *tree,* one person might visualize a huge, spreading maple tree, another might think of a Christmas tree, and a third might picture a scrub oak.

7. Less effort is required to understand visuals. With text your reader is forced to follow the writer's sequence of words, thoughts, and the manner in which the writer composed them. With a visual, a reader can start anywhere he or she likes.

8. Visuals save words.

SYNERGISM

When visuals are used effectively, along with the proper text, the phenomenon of synergism results. *Synergism* is the combined impact of separate inputs that, when effectively combined, increase each other's effectiveness. The total effect is greater than the sum of their individual effects; in other words one plus one equals three in this case. Your material can be three times more effective when visuals and text are properly combined. To achieve synergism, the visuals must be tied to the text so the two complement each other. If they don't, the theory "a picture is worth a thousand words" does not apply.

GENERAL CONSIDERATIONS

Once you have your article fairly well outlined, it's time to plan your visuals. Although the number of visuals can vary considerably from one type of article to another, a rough guideline is to use two visuals for each 1,000-word periodical page. Each visual typically replaces about 150 words of text, resulting in about 700 words of text and two illustrations per periodical page.

Note that all visuals, except tables, are generally assigned consecutive figure numbers and are referenced in the texts.

Visuals for technical periodicals are printed in widths ranging from about 2 inches to about 6 inches. Most visuals will be only one column wide. Publications use either two- or three-column widths. A three-column width is about 2 inches; a two-column width is about 3¼ inches. So make sure the visuals you choose will still be legible if they are reduced to 2 to 3¼ inches in width.

Each visual should carry only one idea, in simple form. Complex visuals defeat their own purpose. Each visual must also have a specific title that tells what the illustration is all about. But don't use more than ten words in your title.

Avoid color visuals. They're expensive to compose and convert for use, and most technical periodicals refuse to use them.

Later, when you're writing the text of your article and are unable to describe something, ask yourself, "Could I do this better with a visual?"

TYPES OF VISUALS

Following are the kinds of visuals commonly used in articles, ranked in order of the acceptability, usefulness, and ability to present the maximum information:

1. Line drawings
2. Exploded views
3. Photographs
4. Gazintas
5. Drawing trees, Organizational charts, Flowcharts
6. Graphs, Curves
7. Bar charts, Pie charts
8. Tables
9. Cartoons

The applications, advantages, and disadvantages of each of these visual types will be covered in the sections that follow.

Line Drawings

A line drawing is simply an artist's concept of an object constructed with lines, as shown in FIG. 6-1. The line drawing is the most useful of all visuals when you have to show how an object looks, what its innards are composed of, how it works, etc.

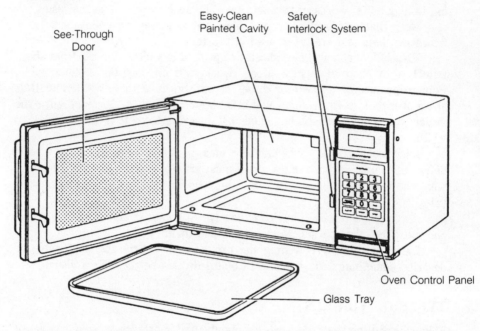

Fig. 6-1. Line drawing.

Line drawings can be restricted to include only what the originator wants to, omitting extraneous details. Cutaways can show the inner workings of the device.

Although line drawings are expensive to make, they can be created from engineering drawings, from the object itself, from photos of the object, or completely from the artist's creative imagination. Also, a wide variety of artist's aids are available from art stores and large stationery stores to help create the simpler visuals. Such items as templates, paste-down lettering, graphic material, and stencils considerably simplify the construction of line drawings.

If you intend to use line drawings in your article, however, check with your editor first. It's possible that the periodical does not want to invest the money required for a complex line drawing, and you may have to supply camera-ready artwork for such an illustration.

Exploded Views

Exploded views are so-called because they appear to be a stop-action photograph taken of an object when it is exploding apart. As you can see from FIG. 6-2, the exploded view is an excellent visual when you need to illustrate the inner workings of an object. However, exploded views are complex, and very costly to draw. Exploded views can illustrate the basic components that make up an assembly and can show how something is assembled or disassembled.

Photographs

When it comes to illustrating precisely how an object looks, nothing is better than the realism of a photograph, as depicted in FIG. 6-3. Photographs show exact details, but this feature could be a disadvantage if there are some components you don't want to show. Photographs give much greater credibility than a drawing because the reader knows the object being photographed actually exists and was not constructed from some artist's imagination. Photos provide drama and informational impact. They usually cost much less than a line drawing or an exploded view, too; however, they can only show surface views of objects. They can't show parts not visible from the exterior as an exploded view can.

To be photographed, an object must be available and in a position where it can be photographed, either in a studio or on site. Photos can be touched up to eliminate some clutter and extraneous information, but only to a limited extent.

Polaroid photos and photos taken with a low-quality camera are marginal for publication. Most editors refuse to use them. Periodicals generally prefer 8-X-10 glossies. They must be mailed in a protective envelope since a creased photo is useless. Never use paper clips on photographs because they can damage the glossy surface. Some editors will accept negatives; however, you may not be able to get them back.

To illustrate size on a photograph, use a ruler, a hand, or another familiar object in the photo. Make sure all photos are sharply focused, are properly cropped to get rid of distracting objects, and have plain, uncluttered backgrounds that do not draw attention from the object you're depicting. Because most visuals in an article are only one or two columns wide, don't expect to use up too much of a periodical page for your photos.

Gazintas

Gazintas, derived from the expression "goes into," are visuals that show the vertical and/or horizontal hierarchy of an object, an idea, an article, an organization, etc. The two types of drawings included in the Gazinta category are drawing trees, and block diagrams.

Ref. No.	Part No.	Part Name
1	590420A	Starter, Rewind
2	590409A	Screw, Retainer
3	590410	Retainer, R.H.
4	590411	Spring, Brake
5	590148	Dog, Starter
6	590412	Spring, R.H. dog
7	590413A	Pulley
8	590414	Spring & Keeper Assy.
9	590536	Housing Assy., Starter (Incl. No. 1)
10	590636	Rope, Starter
11	590387	Handle Assy., Starter
12	590549	Pin, Centering
	34986	Stop & Staple, Rope (Not used on all models)

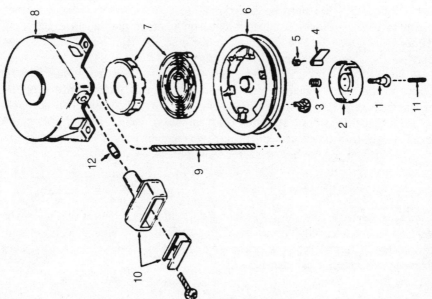

Fig. 6-2. Exploded view.

(COURTESY HEWLETT-PACKARD)

Fig. 6-3. Photograph.

Drawing Trees

Drawing trees, as represented in FIG. 6-4, show the subassemblies that make up an assembly and the assemblies that make up an object. (A company's organizational chart is a form of drawing tree.) The tree in FIG. 6-4 is expanded only partially for purposes of illustration.

This house (top level) is composed of walls, a foundation, a roof, windows, and doors. The walls (Level 1), in turn, are subdivided into their Level 2 parts; studs, insulation, stucco, and siding.

In a drawing tree, all subassemblies of the same relative importance (or size) are on the same vertical level. As you progress from the bottom to the top of the tree, the subassemblies become larger, more complex, and fewer in number, until finally the top assembly, "house," is a single entity composed of all the subassemblies listed below it.

You can use this same technique to diagram the components of a jet plane, a book, or an article. A drawing tree gives a quick representation of the parts

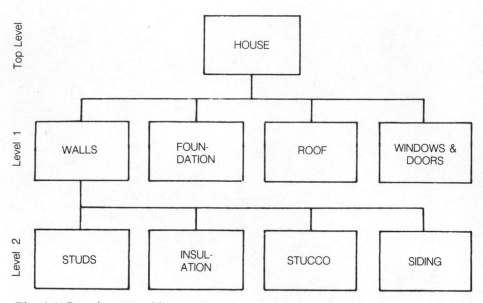

Fig. 6-4. Drawing tree of house.

that make up the whole, their relationship to each other, and the way they are assembled.

Block Diagrams

Another important form of the gazinta is the block diagram. In engineering and science, as in life, most things are not static — they move. From a speeding jet plane to a microcomputer, to a complex chemical reaction, most engineering and science is based on *dynamics,* or objects and materials in action. The block diagram is one of the most often used visuals to show how things react with each other.

Figure 6-5 is a simplified block diagram of an automobile. Turning the key (dotted line denotes an electrical connection) activates a relay on the starter, which in turn applies full battery voltage (dotted line for electrical interconnection) to the starter, rotating the starter motor. The starter is mechanically connected to the engine (a solid line indicates a mechanical link) and cranks the engine. The engine pumps gas from the gas tank via a gas line (double solid line indicates a pipe), igniting the gas, and the engine begins to run.

The engine, mechanically connected to the alternator, rotates the alternator. The alternator is electrically connected to the battery and begins to recharge the battery. Simultaneously the engine is mechanically connected to the transmission. When the transmission is activated, it engages the wheels, which are then turned and the car moves forward.

The block diagram is a very versatile visual for showing the interactions

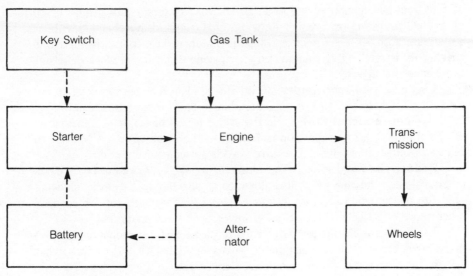

Fig. 6-5. Block diagram of an automobile.

among a number of items, whether these items are electrical, mechanical, chemical, or a combination of the three.

A pictorial form that is easy to comprehend, the block diagram shows how things interact, without showing the details of the individual components. For example, FIG. 6-5 does not show the details of how the starter uses the electrical power it receives from the battery. At this level of detail, it's necessary only to show the overall operation of the starter. Lower-level block diagrams would show the details of how the starter functions. It's confusing if you mix different levels of detail in a given block diagram.

One very important advantage of using block diagrams in your articles is that they are very easy to draw. Anyone with a straightedge can draw a presentable block diagram, and many of the simpler ones can be constructed using a word processor. Block diagrams are a powerful visual for explaining how things react, both internally and to external forces.

Guidelines for Block Diagrams

Over a period of twenty-five years, I have created hundreds of block diagrams and consider them the most useful of all visuals for any type of technical writing. Based on this experience, I have developed a number of rules and guidelines to use when creating block diagrams. They are:

Limit the number of blocks   Use no more than 8 to 10 blocks in any single block diagram. Nothing can overwhelm and discourage your reader more than to be suddenly confronted with a maze of 20 to 50 blocks on one diagram, all with criss-crossing lines. The mind simply cannot grasp such a huge amount of confusing information at one time. Break your diagrams into groups of 10 blocks or less, giving your reader drawings that can be comprehended without a migraine. Also, since your editor will undoubtedly reduce your drawing to one that will fit in one or two columns on a magazine page simple block diagrams are a must for articles.

Use 8½-X-11-inch paper   Use 8½-X-11-inch paper or smaller because it disciplines you to use no more than 8 to 10 blocks in a diagram.

Use conventional flow   Use the conventional direction of information or signal flow of left to right, top to bottom — the same direction you use to read all printed information. Use arrowheads to indicate direction.

Reduce the Fog factor   To reduce the fog factor, or confusion, in block diagrams, keep the number of lines down to an absolute minimum by showing only the major actions and interactions. Detailed interfaces can be covered much more effectively in tables, or in lower-level block diagrams.

Use functional names   Give each block a short, functional name of no more than three or four meaningful words — a label that describes what specific function(s) it performs. Don't use nondescriptive titles such as Valve A. Call it e.g., Flow Control Valve, so its function is easily understood.

Use proper labels   It's extremely important to make sure that the inputs and outputs that interact between separate block diagrams use precisely the same name. It's very confusing if you designate a flow input as Purge Air in one diagram and Cleanout Air in another. Your reader shouldn't have to stop and try to figure out if they're the same. Use the exact, same name for both inputs, and use the exact same name when you refer to them in the accompanying text.

Don't mix Levels   Keep the functional level the same within a specific block diagram. Your reader's mind has been geared to think along a specific functional level for each diagram and it's confusing if you indicate, for example, a spark plug on an overall diagram of an entire automobile.

Alternate linkages/interfaces   Indicate the primary flow with solid lines, and secondary flow within the same diagram with another code, such as dashed lines. Use different types of coded lines to distinguish other types of flow.

Text Guidelines

When you write the text that accompanies the block diagrams, the following guidelines will help:

Use Identical names   Use the identical names in the text that you use in your block diagrams. I'm a fanatic about this because I've found it to be one of the single most confusing deficiencies in technical writing. When you call a block a CPU one time in the text, a computer the next time, and a processor the third time, you're planting unnecessary confusion. Your reader has a difficult enough time trying to comprehend technical concepts without having to pause and try to figure out if the writer is referring to the same part of a computer. The same guideline holds when you label input and output flow from one diagram to another. Designate them by the same name in the accompanying text to avoid confusion.

Define Terms   Define abbreviations and terms the first time you use them. For example, "The *R*andom *A*ccess *M*emory (RAM), provides storage for lost bits."

Follow the Flow Direction   When you write the text that describes the functions of the blocks, use an order of presentation that follows the direction of flow. First describe the input, then how the various blocks massage and control this flow. Carry the flow from left to right, all the way through until it is finally terminated, stored, or output.

Graphs

If you need to show how one or more continuous (or extrapolated) variable(s) change with respect to another, or if you need to show trends or minimums/ maximums, you can best illustrate them with a graph.

A graph makes it much easier for the reader to grasp information than if it were printed in a table. A graph is a picture, rather than a set of numbers. Most graphs are two-dimensional, having a horizontal, or X-, axis and a vertical, or Y-, axis. Three-dimensional (X-Y-Z) graphs are used occasionally to illustrate more complex relationships.

Line graphs are accurate, but their data-presentation capability isn't as precise as for tables. However, communicating trends, relationships, etc. is often more important than showing exact data points.

To emphasize a curve or trend, choose a vertical scale such that the slope of the curve will be 45 degrees or greater. The idea of movement or trend is best emphasized by the steepness of a line and minimized by the flatness of a line. If your graph contains only one curve, you can add another dimension by coding or cross-hatching the area under the curve. If you use more than three lines, however, the graph can become cluttered and confusing and difficult to interpret.

When constructing a graph, label all coordinates simply. Use a minimum number of grid lines and plotted lines. Label curves directly, rather than using a key. Don't run the curves to the ends of the grids; leave a 10 to 15 percent space all around.

Line Graphs

One of the most popular of all graphs is the line graph illustrated in FIG. 6-6. It's ideal for showing a dramatic side-by-side comparison of two variables: the independent on the X-axis, and the dependent on the Y-axis.

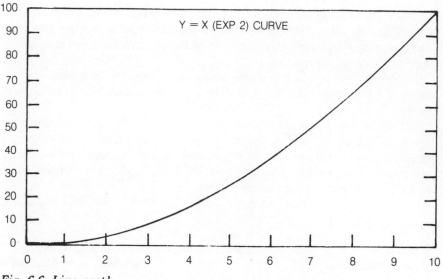

Fig. 6-6. Line graph.

Pie Charts

Pie charts provide an easy-to-understand and dramatic picture of the parts that constitute the whole, as depicted in FIG. 6-7. A pie chart gives a quick look and comparison of an item's parts. It shows proportions and the relative size of related quantities. But a pie chart is not accurate; you can't tell the difference between 29 percent and 32 percent, for example. If accuracy is required, label the percentages on or near each slice.

Don't cut the pie up into more than six slices, though. Additionally, make the smallest slice of the pie greater than 2 percent to maintain effectiveness.

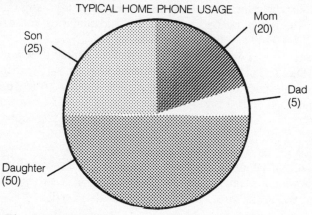

Fig. 6-7. Pie chart.

For emphasis, you can shade individual elements, or cross-hatch them to code the various parts. Make sure you label the various slices with their categories, as well as the percentages of the whole they represent. Also check to make sure all of the slices add up to 100 percent.

Bar Charts

Bar charts are a pictorial method of displaying tabular data when no mathematical relationship exists among the variables. These charts are particularly useful when you need to dramatically illustrate relationships between different sets of data. They give a quick picture and are easier to make comparisons with than pie charts. However, if the bars are too short or too long, they lose their effectiveness.

As illustrated in FIG. 6-8, it's best to separate the individual bars for clarity.

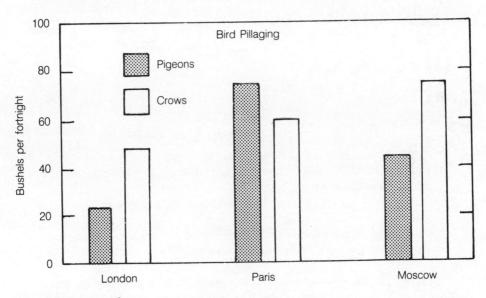

Fig. 6-8. Bar graph.

Use bars of the same width and the same spacing so they are attractive and easy to understand at first glance. You can plot the bars vertically or horizontally. Use vertical bars if you're plotting the type of data that people expect to see vertically, such as temperature, weight, etc. Horizontal bars are suitable for distance, time, speed, etc. You can subdivide the bars by shading or cross-hatching to add another dimension of information (FIG. 6-9).

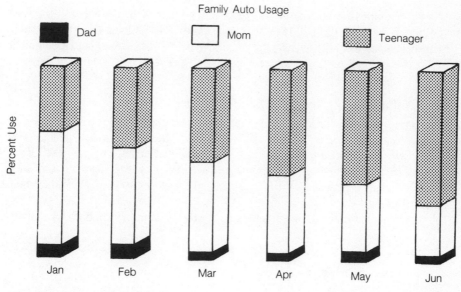

Fig. 6-9. Enhanced bar graph.

Tables

Use a table when:

- It will make a point more clearly than text can.
- You want to discuss the data in the table and draw conclusions from this data in the accompanying text.
- Putting the data in the text would take at least three times as much page space.
- The data in the table should not be duplicated in the accompanying text.
- The data in the table must also be accurate.

If there's no other way to present your information or if you have a lot of data to present and you want to show how they relate to each other, you probably can present it better in a graph. A table is less desirable than a graph because your readers must try to formulate a picture in their minds based on the numbers in the table. But if you need to present numerical data with a high degree of accuracy, a graph won't work. Instead, use a table. Make sure you discuss the figures in the table in the accompanying text to help with their interpretation.

Informal Table

An informal table, because it has no lines or ruled forms, can be included in the middle of text. It can be used to list numbers, products, etc. in vertical columns for visual clarity and quick reference. An informal table must be brief and simple. It is not identified by a table number and normally does not have a title. As an example of an informal table:

The percentage concentration of various chemicals in smog is:

Oxygen	4%
Nitrogen	6%
Carbon Dioxide	90%

The data in the informal table can be discussed in the text that follows it.

Formal Tables

A formal table requires a table number, a table name, and a minimum of internal and external lines to simplify reading the data. Columns and subcolumns should have headings that identify the nature of the data. Where needed, include units of measure, such as ounces, length in feet, percentages, etc. in column headings.

The use of a formal table, as shown in TABLE 6-1, eliminates the cumbersome repetition of units (ft/sec/sec), and the tabular arrangement helps your reader compare data easily and quickly. Often your readers won't be able to grasp the significant differences among quantities when they're presented in paragraph form, whereas in an aligned table the differences stand out.

As a guideline, when four or more sets of data are to be presented, it's best to use a formal table.

Table 6-1 Table Format.

Table Name			
		Multiple Column Caption	
Stub Caption	Column Caption	Subcaption	Subcaption
Heading 1	Data	More data	And more
Heading 2	Data	More data	And more
Heading 3	Data	More data	And more
Heading 4	Data	More data	And more

To prepare effective formal tables:

- Leave a lot of white space.
- Align columns on the decimal point.
- Clearly label all columns — don't over abbreviate and don't use too many columns or rows.
- Use decimals, not fractions.
- Try to keep the table width to the width of one column in the periodical you're writing for.

EXERCISES

1. Create and sketch at least two visuals per periodical page for your article.

2. Integrate your visuals by organizing them with the text to be used.

7

Write the Text

Words are the greatest things ever invented.

Sherwood Anderson

After you've completed your outline and your visuals are roughed out, you're ready to write your rough draft of the body of the document. And it should be a *rough* draft because you need not be concerned about the finer points of grammar at this time. For your rough draft, don't be overanxious about the mechanics of writing. Be more concerned with the technical content. Concentrate on getting the facts down, in the proper order.

When you begin writing, you'll discover that a significant advantage of using an outline is that you will be writing about, and concentrating on, only one topic at a time. Your mind need not wander from one topic to another, worrying and wondering what is coming next. You can mentally set aside all the other topics in your article for the moment and focus your total faculties on the one topic you're writing about.

The "meat" of most documents is the body, which fleshes out the skeleton of the outline. In this chapter, we'll cover a method I've developed in writing hundreds of reports, proposals, and technical articles, and three nonfiction books. This technique will take you through the steps of organizing, writing, and revising your documents. After you've tried these proven techniques, you'll never need another. It makes fleshing out the body of a document easy, fast, thorough, efficient, and gratifying.

WORD PROCESSORS

Whether you choose to write longhand, type, dictate, or use a word processing program to compose your article is your decision. If you have a choice, a word processing program is by far the best. Time and again it has been shown that writing with a word processor improves the quality of an author's writing. Also more and more publishers are requiring that articles be submitted on computer diskettes.

A word processor automates your writing, making it much easier. Unlike a typewriter, you don't have to worry about hitting the carriage return at the end of a line because a carriage return is automatic in a word processor. No matter how fast or how slow you type, the word processor can keep up with you. Text is automatically scrolled so you can keep typing and watch the material you input appear on the screen instantly. When you're through inputting your information, you can go back and rearrange sentences or paragraphs, as well as delete or add words, without the necessity of retyping the entire document as you would if you used a typewriter.

Many auxiliary programs are available to help you come up with near-perfect text, including a spelling program, a thesaurus, word counters, and programs to check sentence length and the use and abuse of jargon. They can also check the use of too many passive verbs, the overuse of pretentious words, etc.

Researchers have learned that people who once dreaded writing have become much more positive about writing once they have learned to use a word processor. Another important fact that researchers have discovered is that our short-term memory lasts about five seconds. A word processor can record your ideas much faster than you can input them, so your fingers, and not the computer, are the speed limit in capturing your ideas.

Besides, word processors are the wave of the future. The days of the handwritten draft being handed to a secretary for retyping are nearly gone. Soon, the only typewriters you'll be able to find will be in museums. Before long most communications will be delivered as soft copy via modems, and heavy, multivolumed instruction manuals will be delivered on small, computer diskettes to be displayed on a video screen.

Computer Knowledge Not Needed

You don't have to know anything about computers to use a word processor any more than you need to know the theory of television reception to operate a television set. All you need to learn are a few simple English commands.

When you first start using a word processor and are learning its commands, you're using the rational/logical part of your brain. And when you're using the rational/logical part of your brain, the creative/intuitive part tends to be inhibited.

However, once you have learned the basic commands, you'll find that the operation of a word processor is *transparent.* In other words, you can type and perform all the necessary commands and not be conscious of it, just as you do when you drive an automobile or walk. This is the point of reward, when the operation of the word processor becomes automatic and your creative/intuitive function begins to assert itself. Once you reach this stage, you are truly accomplishing creative writing. This is the magic time when writing becomes exciting, a pleasure.

Some important things to remember when you're using a word processor:

- When you start feeling comfortable using a word processor, it becomes very easy to stop and make corrections when you're inputting. However, when you're doing your creative drafts, strongly resist this impulse. When writing creatively you are only concerned with ideas, not grammar or spelling. All you should do in the early stages of writing is get your ideas down, and not be restricted by matters of form that can be attended to during rewrites.
- If your word processor doesn't have an automatic Save feature every few minutes, be sure to perform one yourself. Pause in your writing every ten to fifteen minutes and save what you've written, just in case you lose power.

Word Processor Capabilities

The word processor program you use should have the following capabilities as a minimum:

Help menu   A Help menu should be available at all times, regardless of the writing mode you're in, and should summarize the major functions of all of the key commands.

Insert mode   The Insert mode lets you add information in any section of a document. The remainder of the material moves to the right to make room for the added information. This capability lets you start with any topic you choose and add any amount of information, in any part of your document, and lets you repeatedly switch back and forth between topics and still keep everything in order.

Delete   You should be able to delete a letter, delete a word, delete to the end of a line, and delete entire blocks of text.

Block Move   The Block Move command is used to move large sections of text from one location to another. It is often referred to as *electronic cut and paste.*

Search   The Search command locates any part of your writing for revision and correction. You should be able to search *globally;* that is, search for a specific phrase throughout the entire document.

Search and replace   This command allows you to correct spelling errors or to change words in your text. You should also be able to search and replace phrases globally.

Underline   The Underline command enables you to highlight words, phrases, and titles in your article.

Boldface Boldface, like Underline, enables you to highlight words and phrases. Use it for headings and subheadings.

Tab Set Tab Set is used for setting indents, table formats, etc.

Centering This command enables you to center figure titles, tables, etc.

Right Justify Use this command if you need even left and right margins on your printouts.

Line Length This command enables you to set the line length to meet the special requirements of periodical publishers.

Graphics Some of the more advanced word processors have a Graphics feature, which enables you to draw simple graphics (block diagrams, flow-charts, tables, etc.) on the screen and print them out using a dot-matrix or laser printer.

Print The Print command enables you to output your text on paper. Usually a dot-matrix printer operating in a *Near Letter Quality* (NLQ) mode is acceptable. Some periodicals prefer letter-quality printers. For graphics, a laser printer is desirable.

Print Spooling With Print Spooling, you can print one document when you're working on a different document. This feature is highly desirable.

Cancel Print You should be able to cancel a print in progress if you make a mistake or change your mind.

Disk Storage Some periodicals might require that your article be submitted on a diskette, in addition to a printed copy, to save the publisher the added burden of manually inputting your prose into their computer.

It would be nice if your word processor also had these features:

Speller A spelling program is handy to correct your misspelled words. The speller should also be able to catch double words and to add words to a special dictionary to store some of the jargon of your profession. Your speller should be able to correct words *in context;* that is, as they appear in your text.

Thesaurus A thesaurus helps you find synonyms for words you might tend to repeat often.

Macros Macros provide shortcuts, such as designating a special key to type in a complicated word, to set up a format, etc. A macro is a simple computer command that combines and reduces a word processor command requiring a number of programmed strokes to one or two key strokes.

Word Count Some spellers have the ability to count the words in your text. This saves a lot of work.

Diskette Storage Capability

One "byte" represents a single character — it might be a letter, a symbol, or a space. An average word length is 6 characters, so a 150K (150,000-byte) diskette can hold about 25,000 words. An average double-spaced page contains about 250 words, so even a low-capacity diskette can hold 100 pages of writing. You'll never have to worry about running out of storage space on a new diskette when you're writing an article.

MECHANICS OF WRITING THE ROUGH DRAFT

The first step in drafting an article is to gather all your index cards, notes, lab notebook, data sheets, and visuals. It's very important to make a concerted effort and spend an adequate amount of time planning what you're going to write, rather than to just sit down and write what comes into your head.

Not only is an unplanned article difficult to write, the result is often a rambling, confusing document that few people will bother to read. Also, when you start writing, you'll find that the words tend to set hard, as if in concrete. Once written down they are difficult to edit, rearrange, add to, or modify your article.

Before you start the actual writing, plan your article in a considerable amount of detail. Once planned, writing is much easier. The flow of ideas and conclusions will be arranged in a logical fashion.

Know Your Audience

Before you start to write a technical article, remember that you're writing for a specific audience. So, pause a moment and consider your readers. Are they:

- Technically inclined?
- Familiar with the background subject of your article?
- Busy, with little time to read long articles?
- Of varying backgrounds — some scientific, some executive?

Remember, your reader is not forced to read your article. It's up to you to organize and write it so it's interesting enough to be read.

When you write an article about something you've accomplished, it's only natural that, since you've been deeply immersed in your work for so long, you'll want to write about every little detail. But don't assume your readers will be interested in all these details. They need only enough information to answer the questions, "What is this all about?" "Why should I read this?" Readers will read for the article's contents, not for the joy of reading your prose.

Follow the steps listed here in writing the body of your article. It's a proven formula that will help you come up with a well-written article with a minimum of effort.

List Your Topics

To start, look over all your file card notes and visuals and review the working outline you made in chapter 5. Find a quiet place where you won't be disturbed.

Write the Text

Pick a time of day when your mental equipment works the best. If you're a morning bird, like me, choose a nice, quiet morning to work. Use a deserted office, a quiet corner of the company library or lunchroom, or an empty conference room. Sit down and concentrate. Review, read, and reread your index card notes and the other notes you've accumulated. Study your working outline in great detail. Use earplugs if you need them.

This is the single most important step in the preparation of your document, so spend at least a few hours reading, reviewing, questioning, checking, and double-checking your references and your notes. You want this information to sink deeply into your mind.

Take a Break

The next step is also very important. Put your notes away and do something else for a few hours, something totally unrelated to your report. This gestation period is needed so that the material you just studied can settle deep in your mind and that marvelous mental mechanism, your subconscious, can go to work on the material to sort, evaluate, classify, organize, expand . . .

It's ideal if you can use this time to engage in some sport or physical activity, such as jogging, bicycling, hitting a golf or tennis ball, swimming, etc. Don't engage in any work that requires deep concentration or any taxing exercise of your mentals (except for, perhaps, a crossword puzzle). And don't even consciously think about your article. Put it out of your conscious mind and concentrate on a physical activity, talk to friends, or go to a movie.

Just relax for a few hours.

Write As Fast As You Can

After you've been away from your material for a few hours, find another quiet place and get your equipment ready. Sit down in front of your word processor, your typewriter, or your dictating machine, or get a supply of pencils or pens and loose-leaf paper.

Take out your working outline if you typed it on a separate sheet and tack it on the wall. Or, if you're using a word processor, display your outline on the video screen. Also tack up the visuals you're going to use so you can refer to them. Only your visuals and outline should be visible now.

Don't consult your detailed notes under any circumstances at this point! This is very important! Seal your detailed note cards in an envelope if you must, but do not look at them now. The problem with writing directly from notes is that you'll end up simply copying the notes directly into your text. You'll be using the exact words from your notes, instead of putting all your thoughts and ideas into your own words. Working directly from notes stifles creative writing. It's better to put your creativity to work and write using only your working outline and visuals to guide the words from your subconscious mind to the keys.

Next, using either your typed working outline or the outline you input in your word processor, review the topics and decide which of those listed in your working outline will be the easiest to write first. Then start writing whatever comes into your head, referring to the applicable visuals as you write. This way you'll be writing a series of short, easily handled topics, rather than one long, continuous one.

Write or type as fast as you can. This is important! Don't let anything slow down your creative process. Double- or triple-space your writing or typing. You'll need this space later for corrections, additions, etc. Don't pay any attention to grammar, spelling, sentence structure, variety, or length. Use incomplete sentences if you must. Just get the words and ideas down.

Thoughts are associative. One thought leads to another. You must allow your thoughts to flow freely and use the style and vocabulary that are natural to you.

If you're stuck for a technical detail, a name, or a number, leave a blank space or a question mark. You can fill in the details later.

Don't let anything sidetrack you in this highly creative process. You'll find that your enthusiasm will grow once you get over the initial hurdle of starting. You'll be amazed and pleased at how quickly thoughts come to you. Your hands will not be able to keep up with your brain. Words, thoughts, and sentences will pour out of you in a logical sequence, forming a smooth-flowing narrative that you'll be proud to have authored.

As you finish each topic or subtopic, cross it off your outline. If you're writing or typing, use a new sheet of paper whenever you switch to the next topic. Write the topic name and page number in the corner as a precaution in case your sheets get mixed up. Here again the beauty of the word processor is demonstrated, because all you need to do is switch to any part of your topic outline and continue writing. All the text will simply move over to make room for what you are adding.

If you get stuck on a particular topic, switch to another one. Chances are that you'll soon get unstuck and be able to go back and finish that topic later.

The subconscious mind is probably one of the most fantastic creations in existence. It makes a supercomputer look like a windup toy. You'll be amazed at how easily you can turn out good writing, fast and almost effortlessly, with your subconscious mind running in high gear.

Check Against Your Detailed Notes

Soon after you've finished writing your rough draft, unseal the envelope with your detailed notes in it. Read through your notes again to see if you've left anything out. You probably have, so fill in the blanks left open in your important first draft.

If you're writing by hand or typing and you have left out quite a bit so you have to add a page between pages 2 and 3, number it 2a. That's the beauty of using individual sheets of paper: you can easily add, subtract, or reorganize as required. In a word processor you can easily insert whatever information you left out and whatever corrections you need to make. If you're dictating, you'll have to name the topics you are modifying and input your comments verbally or you'll have to have a rough draft typed and then mark it up.

Review

Before you start the next review, pause for a few hours (or a few days if your deadline permits it) to let your writing cool off. "Write in haste, polish at leisure" is good advice.

Then, put on your editorial hat and go through what you've written, from beginning to end. Check the sentence length, structure, grammar, and punctuation. Pause to check any spelling you're unsure of. If your word processor has a spelling program, use it. The speller will catch many of the mistakes you'd miss on a visual check.

I find that writing of things technical (and writing in general) takes two distinct and separate mental processes. First you write creatively, using the creative/intuitive part of your brain and concentrating on getting the proper technical information down, in the correct order, in your own individual style. Next, you put the logical/rational part of your brain to work and become an editor, critically blue-penciling as you read your material objectively, checking for proper grammar, spelling errors, etc. You cannot accomplish these two separate and distinct processes at the same time because, if you edit when you write your initial draft, your creative processes will be stifled. Your mind will have to continually shift gears back and forth. This mind-jerking strips your mental gears and ruins the writing process because creativity must exist without any bounds or rules. Creativity can't be tied down or restricted by grammar.

When you have your technical writing hat on, you're doing some pretty heavy creative writing. You're writing what has probably never been written before. It's coming out of your unique mind and can only be written by you.

Editing, however, is less of a creative task, but it is a vital one. Editing applies certain established, well-proven rules that have become accepted over the years. It's an extremely important task because misspelled words, awkward sentence structure, and improper grammar will completely distract from, and distort, the important things you're trying to say. Whenever you read a document replete with misspellings and errors, your mind is sidetracked and you begin to lose interest and confidence in the accuracy of the writer's technical information.

For this step, read your article through, from beginning to end, blue pencil (or pen) in hand, and make any needed corrections you find. If you're working with a word processor, you can accomplish the process on the video screen by scrolling through the text or using a printed copy. Write notes to yourself on the screen or on the printed copy about what to do later. Don't interrupt your reading too long during this process. However, read the entire draft through, at a leisurely pace.

When you come to a sentence you can't understand, just imagine how difficult it would be to understand for someone who knows little or nothing about your topic and who is reading it for the first time. Make a note to clarify it, but don't stop to correct it at this time. Wait until the next step to make the necessary revisions.

Revise

A few hours, or a day or so later, take the necessary time to do the detailed rewrite and make the additions and corrections required. Put the spelling checker to work if you have one. Use the electronic or manual Cut-and-Paste function to restructure your article as you noted during your review. Because second and third thoughts are often clearer than first thoughts, revision can be the difference between an excellent article and a mediocre one.

And revision is much easier when you're word processing. If you're working with a word processor, it's best to perform this revision step with a printout of your entire article because all of your errors will show up better on paper and you can easily jump back and forth to check and verify things in a printout.

Review the Article

Set your article aside for a few more days. Then read it through, from beginning to end, and make another set of corrections.

Check your sentence lengths. Your average sentence length should be less than 20 words; 17 is ideal. For a quick visual check, a typewritten copy usually has 10 to 12 words per line, so your average sentence length should be less than two typewritten lines. Check for long words; they are usually abstract words. Replace them with concrete words whenever possible.

BEAUTIFY YOUR BODY

Now that you have your document drafted, it's time to beautify the body: to add bullets, listings, headings, etc., to make your document easier to read and understand. Experts agree that graphic relief is needed to create white space and to make large blocks of text more readable and more attractive. I call these graphic reliefs *literary cosmetics*.

Typical literary cosmetics you can use are:

- Paragraphs
- Bullets

- Listings
- Headings/Subheadings
- Underlines

Paragraphs

Go through your material and check that your paragraphs are not too long. Keep most of them under about ten lines on a typewritten page (that's about 100 words). A final typed page should have at least three or four paragraphs per page.

But don't overdo it. Too many short paragraphs distort the effect you're trying to create: lots of white space, variety, a break for the reader's eyes.

A reader sees a paragraph as a certain amount of information he or she must gulp down in one swallow, so make them small swallows. Short paragraphs throw more *light* or understanding on your page, making it easier to read. Paragraphs also help your reader group and understand topics better.

Bullets

Bullets are the symbols used to set a list of information off from the text to emphasize certain points you're going to make. For example:

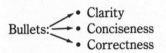

Bullets are an additional effective way to give the reader's eyes and mind a break when you enumerate and emphasize certain items. But don't overdo them, either, or they will lose their effectiveness.

Numbered Listings

A listing uses numbers or letters to list things in sequence and set them off for the reader as individual, important topics. For example:

1. Grammar
2. Sentence structure
3. Spelling

A listing is a quick way to summarize points for your reader. It makes the items stand out. Lists take more space than sentences, but they do emphasize the data they present.

Headings

Headings provide a visible outline of your article. They also form a summary of, and introduce, the material below them.

Headings show the subjects you're covering and the order in which you're covering them. A reader can obtain a quick summary of a properly written article by reading the headings; noting the order in which they're arranged and the amount of space allocated to each. Headings can whet a reader's appetite and entice him or her into reading your entire article.

You need about one heading per two typewritten pages of 500 words to allow more light on the page. Headings should be less than five or six words long; two- or three-word headings are common. When your article is published, your publisher will usually convert your headings into boldface type so they'll stand out even more.

Another advantage of headings is that they save creating a lot of transitional phrases by using headings to bridge two topics coming up next. A heading helps you jump from one aspect of a subject to another without such much-abused transitions as:

- On the other hand . . .
- Meanwhile, back at the ranch . . .

Underlining

Underlining is used in the text of an article to call attention to a specific word or phase you want to emphasize. You're calling the reader's attention to these words.

But like most good things, underlining can also be overdone. Use them judiciously.

EXERCISES

1. Draft your entire article.
2. Review and revise your article at least three times.

8

Prepare Your Manuscript

*A whole is that which has
a beginning, middle, and end.*

Aristotle

Now that you have a well-formed body, it's time to put a head (lead), a foot (conclusion), a hat (Title), and a kerchief (subtitle) on your masterpiece for your readers. Each of these sections forms a significant part of your article. The relative attention-catching ability of each of these sections and the order in which the sections are read are illustrated in FIG. 8-1. The larger the area of the rectangle, the greater the stopping power — the greater the ability to arrest and hold a reader's eye, to pique your reader's interest.

As stated earlier in this book, the rising degrees of involvement for your article occur when the reader:

- Stops to read the title and subtitle. HEY! YOU!
- Looks at your visuals. HEY! YOU!
- Reads the first paragraph. HEY! YOU!
- Reads the body. SEE?
- Reads the conclusion. SO!

TITLE

If you haven't already chosen a good title, now is the time to work on it and select one that will grab the reader. The title, or *headline,* in huge, bold print is the first thing most potential readers will see. It must be precise and clear, telling them what they are about to invest their valuable time in. If the title doesn't hook them, they'll simply turn the page and your article will go unread.

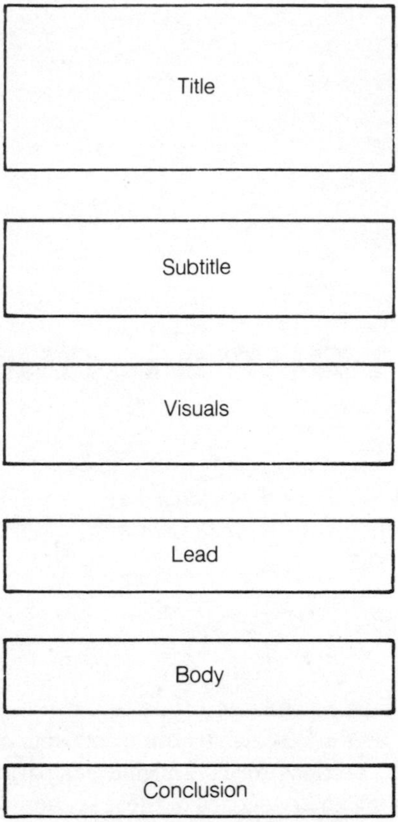

Fig. 8-1. Attention-getting values of parts of an article.

If you're not satisfied that your title will entice readers into your article, review the advice given in chapter 2. Come up with a number of possible titles (use no more than about eight to ten hard-hitting words) and work until you get a good one, an attention-grabber.

In one of my articles, the title was "Potent Visual Aids Get Your Message Across," in bold, black, half-inch letters. It told the reader how the article would help him or her use visual aids effectively. The title summarized the article in only seven words.

Once your readers have read the title and it has generated some interest, they'll read the *subtitle*. A subtitle is an expansion or clarification of the title, often in bold print slightly smaller in size than the title. A subtitle can be a sentence or two, also in bold print (typically ¼-inch letters), such as was used for my article.

Nervous about Making That Required Speech?
Visual Aids Not Only Focus Attention Away from You;

They Make for a More Effective Oral Presentation.
But Be Sure to Create and Use Them Properly.

Note also how the HEY! and YOU! parts of the four-part formula have been used. The title is basically a HEY! component that quickly and effectively informs the reader of the subject of the article. The subtitle uses a mixture of HEY! and YOU! and even includes the important word *you* in the subtitle.

Next, using your title as your guide, write a short sentence or two to use as a subtitle, expanding on the theme of your article and involving the reader, the *you*, in your subtitle.

VISUALS

If your title and subtitle have intrigued your reader, he or she will next glance at your visuals. Usually more people will glance at your visuals than will read your text. For this reason it's important to write the body so that a visual is included on the first page of your article. Review your visuals and make sure an interesting one is incorporated in the first 500 words of the body.

LEAD

Next, pull your readers into the body of your article using the *lead*. Some commonly used leads follow:

- The *summary* lead tells succinctly what the article is about. This popular type of lead is the easiest to write. It might be one or two paragraphs in length and list some of the main facts in your article. This type of lead can also serve as an abstract of your article.
- The *question* lead piques the reader's curiosity by asking a question the reader would like to know the answer to.
- The *problem/solution* lead is appropriate when your article discusses the solution of a basic problem.
- A *description* lead can describe in the first sentence the object, technique, or process your article covers. You should be able to use words that evoke concrete images for this type of lead.
- A *comparison* lead is ideal when you are analyzing or comparing two or more processes or equipment.
- The *news* lead is useful when you're announcing or describing the first installation of a unit, a new product, or a new process.
- In a *statistical* lead, you begin by comparing statistics that are shocking or very different from the norm.
- An *anecdotal* lead begins with a mini-story concerning people, and will illustrate and introduce the topic of the article. This important, human-interest lead is, regrettably, used very little in technical journals. (I used it in this book in chapters 1 and 5.)

- A *historical* lead begins by giving the origin or background of a product, service, or discovery.

A lead performs three basic functions:

- It serves as a transition from the title and subtitle into the body of your article.
- It summarizes your article and hooks your readers, promising them more information of interest to them if they'll continue to read.
- It shows them how they will directly benefit from reading your article.

As an example of an opening paragraph, consider this one from my article on visual aids:

Sooner or later you're going to have to give a talk, perhaps at a program-review meeting or a technical conference, in a discussion with a customer on a proposal's technical aspects, or as part of a presentation to management. And, when you do, you'll discover that visual aids, in conjunction with your oral presentation:

- Provide an outline of your talk for you and your audience to follow.
- Help keep the audience's attention.

This first paragraph pulls the readers into the article by reminding them that sooner or later they're going to have to give a speech. This will appeal to almost everyone involved with any kind of technical work. Then it jumps right into the usefulness of visuals in such a presentation. Notice the YOU! component is used again to convince readers that they will benefit from reading the article.

The most important sentence in any article is the first one; it's a do or die situation. If it doesn't do (entice the reader to read your second sentence), your article just died.

The body you wrote in chapter 7 now needs a lead to pull your readers into the main part of your presentation. Write an opening paragraph that expands on your subject, one that complements the title and subtitle and hooks your reader into your article.

THE BODY

If you've done a good job with your title, subtitle, and first paragraph, you should now have the readers hooked into reading the rest of your dissertation, so the body will take care of itself.

CONCLUSION

For the conclusion, you can reword what you said in your first paragraph and leave the reader with some satisfactory thought or call to action. For example, in my visual aids article I concluded with:

> Use a lot of visuals the next time you have to give a speech. You'll find that they will do much of the talking for you.

As you can see, usually one paragraph is enough for a conclusion, but don't be afraid to use more if needed.

A conclusion can make recommendations for further developments of your subject, or it can summarize the results of your research to re-emphasize them. It can also encourage your reader to take some action. So, add a last paragraph to your article that summarizes what it's all about.

MECHANICS OF HARDCOPY SUBMITTAL

When you prepare the final draft of your text for submittal to your publisher by hard (printed) copy:

- Use a good quality of paper, typed on one side, and double-spaced.
- Leave plenty of space all around to frame the article: about one-inch margins on both sides, the top, and the bottom.
- Type your name on the top of each page (if you're using a word processor, add a header specifying your name on each page) and number each page consecutively, as "Page 2 of 8."
- Make sure the draft is neat and error free.
- Use either pica or elite type fonts. Don't use script or fancy font. And use either a laser printer or a near-letter-quality printer. Some editors are opposed to regular dot-matrix printing.

As an example of what a first page should look like, check FIG. 8-2.

For your visuals, follow these guidelines:

- Put each visual on a separate sheet of paper. Most periodicals will redraw them, so they need not be of drafting quality.
- Do not send originals; send a good-quality copy, unless an original is required.
- Make sure all your figure letters and numbers are clear and unambiguous. Define any special terms used.
- Caption each visual with a title and a figure number.
- Use standard drafting practices and standard symbology on your visuals.

Your Name   About XXXX words
Address
City, State, ZIP
Phone Number
Social Security Number

(Leave about a half of the first page blank)
TITLE OF ARTICLE
by
Your Name

Text begins here . . .

Fig. 8-2. Title page format.

Some special precautions to observe when you submit photographs are:

- Glossy 8-X-10-inch photographs are preferred; 5-X-7-inch glossies are acceptable. Mail them using stiff cardboard or some other protector in the envelope.
- Do not fold, paper-clip, or staple photographs.
- Stamp or print *Do Not Bend* on the mailing envelope.
- Write the captions and figure numbers on the backs of the photographs using a self-adhesive label, or use a transparent overlay to label the photo. Don't write on the back with a pen; it might make the photo unusable.

MECHANICS OF SOFTCOPY SUBMITTAL

Although most periodicals will still accept hardcopy submissions (typewritten copy), more and more are encouraging a softcopy (computer diskette or modem transfer) submittal, with a hardcopy. In the future most periodicals will insist on softcopy because it saves them the time and expense of retyping your manuscript. All they have to do is edit your article and format it for their specific printing requirements.

After the periodical has edited and reformatted it, they'll probably send you a diskette or printed copy of their edited version for your review and approval. All of the chapters of this book were delivered on computer diskettes to the publisher, TAB BOOKS.

Another reason that softcopy submittal will be required in the future is so publishers can send your article by modem or on a diskette to data banks. The data banks can then store this electronic manuscript so that it can be accessed by on-line systems, as discussed in chapter 9.

MAGAZINE SURVEY

To help define current and future requirements for softcopy, I queried 75 representative periodicals. The periodicals covered a wide range, from small presses to large presses, and a broad variety of scientific and engineering disciplines. The results of this comprehensive survey follow.

An overwhelming 90 percent of the periodicals either required or preferred that a prospective author query before writing an article. This saves both the editor's and writer's time and effort.

About 60 percent of those replying said they either require now, or will in the future require, that articles be submitted on computer diskettes. Nearly all editors said they still will accept hardcopy in lieu of softcopy, but this will likely change in the future.

The preferred format for softcopy was 5¼-inch diskettes; some replied that 3½-inch diskettes were also acceptable. The overwhelming choice for type of computer was IBM compatible. The ASCII format was most requested, with a scattering of WordStar, WordPerfect and XYWrite acceptable as word processing programs.

To be safe, then, the best bet is to use an IBM compatible and save your article on diskette in ASCII format. Some periodicals stated that submitting an article in a preferred format would result in a larger honorarium for the author.

Pay ranges varied considerably. About one-fourth of those responding do not pay for articles. Most of these were professional society publications, which is understandable. None of the periodicals stated the author would have to pay to be published. The pay rate ranged from $40 to $250 per page; the highest rate quoted was $2,000 for a feature article. Most pay rates were from $50 to $100 per page.

Article lengths ranged from 500 to over 4,000 words, or from about 2 to 16 pages of double-spaced copy. The average length requested was from 1,000 to 2,000 words.

Graphics are as important as text to all of the periodicals, and most of the replies stated that one to two visuals should be used for every two or three typed pages of text.

MAIL IT

Before you mail your article, make a photocopy of your hardcopy and a backup copy of your diskette, just in case something gets lost or damaged in the mail, and so you'll be able to answer any questions your editor has. If you mail a diskette, use one of the special diskette mailers you can purchase in any stationery store. Mail it First Class, with *Do Not Bend* printed on the outside.

WHAT'S NEXT?

Within four to six weeks you should hear if your article has been accepted or rejected. You may get one of three possible replies from your editor:

- "Congratulations. Your article has been accepted and will be published within the next four to six months. About six weeks prior to publication we will send you an edited version of your article for you to review and approve. At that time we will send you a check for $_____. When your article is published, we will send you a copy of that issue, plus an additional copy from which you might wish to order reprints. . . ."
- "We like your article but we feel it needs some revision before we can publish it. We have assigned Hubert Humbert III as your editor, and he will be contacting you about these revisions. . . ."
- "We are sorry to have to inform you that your article does not meet our editorial requirements, so we will not be able to publish it. . . ."

If you receive one of the first two responses, enjoy. You're about to become a published author. And when you get the galley proofs to review (or a computer diskette if you submitted a softcopy), don't try to rewrite the entire article. The periodical has too much invested in editing and illustrating your article to make major changes. Review the proofs for technical accuracy. Leave the editing to the professional editors. After all, they've purchased it and can pretty much do with it what they want.

If you receive the third response, don't despair. You have tried and you are going to try again. Take heart from the appropriate words of Theodore Roosevelt:

"Far better it is to dare mighty things, to win glorious triumphs, even though checkered by failure, than to take rank with those poor spirits who neither enjoy much nor suffer much because they live in the gray twilight that knows not victory or defeat."

If your article has gone through the entire cycle of query, approval, and writing, chances are excellent that another periodical will buy it, so query the next one on your list. Send just the query, not the completed article, to the next periodical. It takes an editor too long to respond to a complete article. Also one editor's format and requirements might be quite different from another's, so you might have to do some rewriting before another periodical will publish it. And don't inform the editor that your article has already been written and rejected.

COPYRIGHT

Most articles are copyrighted by the periodical that publishes them. However, most periodicals will let the copyright revert back to you after the initial publication, if you so request.

EXERCISES

1. Prepare your article for submittal:

    a. Compose a subtitle.

    b. Write a lead.

    c. Write a conclusion.

9

On-Line Systems

Man is a tool-using animal.

Thomas Carlyle

A revolution is underway that will radically change the look and operation of our public and private libraries, the manner in which our newspapers and books exist, and the general way in which we obtain information. This revolution in on-line systems has been brought about by a number of factors.

Information processing is growing so rapidly that it is impossible for a person to keep up with the advances in his or her profession by reading current periodicals and journals. For example, chemical and biological abstracts now exceeds a quarter of a million per year, making a manual search an impossibility. Professionals need help in locating this information quickly and efficiently.

The low cost and wide availability of the personal computer, coupled with the continually decreasing cost of digital storage, have arrived in time to become the solution to this dilemma. Taking advantage of these phenomenal advances, many on-line vendors have created on-line data banks. *On-line* means you can connect with and obtain information from these remotely located databases through the keyboard of your personal computer and via telephone lines. Data banks are the repositories of this information.

ELECTRONIC LIBRARIES

On-line databases (electronic libraries) have met this demand for ways and means of making these searches feasible and fast. Modern on-line system vendors allow you to tap into gigantic databases, search millions of documents in seconds, and view the result on your home or office personal computer. Suddenly any fact you need is as close as your telephone.

As an example, a doctoral literature search that might have taken six months a few years ago can now be done in less than ten minutes, with far more thorough results and much less cost. A manual search can cost from 10 to 100 times as much as a computer search. Also, some manual searches are a physical impossibility because access to the needed books and periodicals is not possible.

As you conduct research for your current and future articles, acquaint yourself with the use of the databases, the electronic libraries that you can access now that are as close as your telephone.

WHAT IS A DATABASE?

A *database* is an organized collection of information that a computer can search as a unit. A database can be a catalog of manufacturer's products, a series of books or periodicals, or a computer-readable version of a set of encyclopedias. It can include virtually any information that can be converted to digital form and stored for retrieval by a computer. Many databases are collections of abstracts and bibliographies from periodicals. Some are specific abstracts collected by specialty organizations. The variety, depth, and diversity of available databases are mind boggling.

Databases are sold to, or generated by, on-line vendors, who in turn sell access time to anyone wanting to search for information. Your computer, your modem, and your telephone give you this access.

Basic reasons for using a database when researching an article are:

- They help you quickly locate those few important bytes of information you need in that megabillion-mountain of data.
- You obtain instantaneous answers.
- Vendor's databases are much larger than public and private libraries.
- They can be accessed from virtually any place that has a telephone.
- Information is up to date.
- Specialized databases exist for virtually every profession.
- Information is available when you want it. You don't have to worry that someone has checked out the book or periodical you need to consult.
- The information is comprehensive, so you can research to any depth you need.
- You can query from your home or office. No longer do you have to waste time commuting, fighting city traffic and parking problems.

EQUIPMENT NEEDED

The basic equipment you need to go on-line and let your fingers do the talking is:

- Personal computer
- Modem
- Communications software program
- Printer
- Telephone

The basic hookup required is shown in FIG. 9-1.

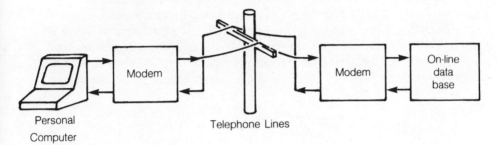

Fig. 9-1. Basic on-line setup.

Personal Computer

As far as types of personal computers are concerned, some vendors' on-line systems will work with all major brands. Many require that you use an IBM or IBM clone. You'll have to check with the specific vendors to see what computer protocols they'll accept.

Whichever computer you are going to use must have a serial RS-232C port; that is, a serial or asynchronous communication card. Some computers come equipped with a serial card; others require that you add one. A color monitor is not required, but if you want to display graphics on your monitor, make sure your computer has that capability.

Modem

A modem must be added to most computer systems. Two basic types of modems are in current use: circuit-board or stand-alone. (A third type — the acoustic modem, which contains rubber cups that clamp over your phone headset — is also available but is rapidly going out of use.)

The circuit-board modem conveniently fits inside your computer. The stand-alone modem is in a separate chassis that connects to your computer via a cable. The basic capabilities of the two types of modems are virtually the same. However, the circuit-board modem does use some power from your computer, and it cannot be transported easily so you can work on a different computer.

The stand-alone modem can easily be moved from one computer to another, but often it is inconvenient to mount and hook into your system. The stand-alone also has LEDs that help you check on the progress of your modem. Stand-alone modems usually cost more than internal modems.

Regardless of the type of modem you use, make sure your modem is, first of all, Hayes compatible. Hayes is the de facto standard for modems; fortunately most modems are Hayes compatible.

As far as speed is concerned, 300, 1200, and 2400 bits per second (bps) are in common use. You often pay a higher price for faster speed, but even if 1200 bps costs twice as much as 300 bps, you can receive data at four times the speed, so it can be a bargain if you are calling long-distance.

When you choose a modem, make sure it has the capability of working at all three speeds. That way you can choose the specific speed that satisfies your requirements, as well as those of the varied vendors. A speaker on the modem card is desirable so you can get some audio feedback on the progress of your call.

Telecommunications Program

Software is needed to tell your computer and modem what to do. A wide variety of telecommunications programs are available, ranging from free to a couple hundred dollars.

Some features to look for in a telecommunications program are:

- *Auto-Dial* The ability to dial a number or series of numbers that the computer has stored in memory, rather than having to dial your number manually.
- *Auto Re-dial* The ability to automatically redial a number if the line is busy.
- *Duplex* Both full-duplex and half-duplex capability.
- *Automated Signon* The ability to send the network number, account number, and password to the host automatically.
- *Break key* The ability to stop the host from whatever its doing and wait for another command. This is very useful when you discover that you've asked for the wrong information.
- *Data Capture* The ability to record on disk the information sent by the host so you can review it at your leisure.
- *Printout* The ability to print out data as it comes into your terminal.
- *Text File Transmission* The ability to type your message off-line, then go on-line to send your message out.
- *Word Processing* A word processing program so you can compose messages.

TYPES OF DATABASES

Two basic types of databases that vendors have available are:

1. *Full text*   This type of database has the entire text of articles for you to read.
2. *Bibliographic*   This type provides bibliographic references and summaries of printed publications. Most databases are of this type, but sometimes they also contain abstracts of the information you're seeking.

Some of the databases are general purpose and cover a wide range of subjects. Others specialize in various fields. A large number of science and technology databases is available.

Representative databases specializing in science and technology are:

- SCI Search   Vast and expensive, but has neither abstracts nor descriptions.
- Compendex   Has over a million records dating back to 1970, with extensive abstracts and a broad coverage of physics, computers, and all fields of engineering. It covers about 4,500 journals.
- National Technical Information Service (NTIS)   The bibliographic database of the U.S. Department of Commerce. It contains technical reports generated by U.S. federal agencies and their contractors. These reports go beyond bibliographic citations; they also contain information on ordering complete copies of the cited reports by mail.
- Computer Database   More than 530 different publications dating back to 1983 cover computer-related hardware and software.
- Inspec   Focuses on physics, applied physics, electronics, electrical engineering, computers, and control engineering. More than 2,300 journals are scanned regularly, and over 330 are completely abstracted.
- Ismec   The mechanical engineering equivalent to Inspec, this database has information dating back to 1973, but it only has abstracts before 1982.
- Engineering Literature Index   Articles from 3,500 worldwide journals.

When you deal with these databases, you can generally obtain a full-text version of the information you're looking for, either as a printout at your terminal or as a photocopy by mail.

ON-LINE COSTS

The costs to use on-line systems vary from about $3 per hour to more than $200 per hour. In addition, most vendors have a one-time sign-up fee. Check with the individual vendors listed in appendix C to determine their requirements.

When you're first learning to search an on-line system, use the lower and less expensive 300 bps rate. After you've acquired some skill, switch to the

1200 or 2400 rates. If your budget is limited, there are some excellent after-hours databases to use: the Knowledge Index in the DIALOG system, Genie, and BRS After Dark have discount rates and an excellent variety of databases.

QUESTIONS TO ASK A POTENTIAL ON-LINE VENDOR

Before you select a vendor, seek answers to these questions:

1. Does the system have abstracts, bibliographies, or full text?
2. What does the service charge? Are there extra charges for 1200 and 2400 bps?
3. Is hardcopy available?
4. Is the coverage comprehensive and up to date?
5. Can you search on a time basis (e.g., 1985–1987)? Can you use Boolean logic in a search to minimize your hits?
6. Does the vendor have local telephone access?

HOW TO GO ON-LINE

Assuming you have your computer on and your telecommunications program installed, then:

1. Set up the proper parameters so you can communicate with the host; for example, set the baud rate, word length, parity, etc.
2. Dial the number, either manually or with the telecommunications program. Your speaker will sound the notes as they are dialed, then you'll get a short ring. Finally your computer will give you a message, for example: 1200 CONNECT to inform you that you have been connected to the remote modem at a baud rate of 1200.
3. Hit your Return or Enter key. If you don't get a response, hit the Return key again.
4. Follow the vendor's log-on procedures. You might have to give a password and account number.
5. Conduct your search and obtain whatever data you need.
6. Log-off using EXIT, BYE, or whatever your host requires.
7. Switch off your modem and go back to working on your article. Incorporate the data you just found.

If you opt to join one of the major services, they usually have training services available on the necessary procedures.

How to Search Once You Are In

You'll generally start your search with a key word, but be careful to make it a key word that is not too broad in its scope so it will minimize the number of "hits" your title will generate. *Hits* refers to the number of possible bibliographic references for the key word you have chosen. If this number is too large — for example, 100, when all you expected was 10 — narrow your search parameters with carefully chosen defining words and syntax.

For example, suppose you want to locate a computer drawing program for the IBM/XT. If you start out with the key word *draw,* you're going to be overwhelmed with hits. So you should narrow it down by specifying as many of the following restrictors as you can:

- Publication title
- Article type
- Author
- Date of publication
- Combination of words, such as *drawing* and *IBM/XT*

Once you've located your reference, some vendors allow you to have the abstract displayed on your screen. Other vendors will permit the entire article to display, and a few vendors will download the article to your computer. Most of the vendors let you order a printed copy that can be mailed to you for a nominal fee.

MAJOR ON-LINE VENDORS

Some of the major on-line vendor services are listed in this section; their addresses and telephone numbers are given in appendix C. Phone numbers and addresses do change, so if you are unable to obtain a response at the number listed, dial the 800-line directory assistance service at 1–800–555–1212.

Because of the lack of space, only a few representative major vendors are listed here. For a thorough coverage of on-line vendors, check one of the following references:

Computer Readable Data Bases by Martha Williams. Chicago, Ill.: American Library Association, 1985. 2 volumes.

Data Base Directory 1984/85– . White Plains, N.Y.: Knowledge Industry Publications (1984–). Annual.

Datapro Directory of On-Line Services. Delran, N.J.: Datapro Research Corporation (1985–). 2 volumes, loose-leaf.

Directory of On-Line Data Bases, v. 1, no. 1– , Call 1979– . Santa Monica, Calif.: Cuadra Associates (1979–). Quarterly.

Some of the largest of the database vendors are listed alphabetically, along with the types of databases they offer.

Bibliographic Retrieval Services (BRS)   Contains over 150 databases, is easy to use, and covers life sciences, medicine, physical sciences, applied sciences, and engineering.

BRS After Dark   A low-cost system available during off hours and contains over 100 databases.

Chemical Abstracts   Provides access to global scientific information by indexing 12,000 journals from 150 countries, plus patents from 27 nations. Their CA file has over 11 million abstracts. Chemical Abstracts specialties are: bioscience and medicine, chemistry, computer science, engineering, environment and energy, materials and construction, math, physics, and patents. Graphics are available from their databases.

CompuServe   Owned by H & R Block, CompuServe offers a variety and broad scope of nearly 500 distinct topics and services. Investment services, reference libraries in medicine, science, and law and other disciplines are available, plus encyclopedia, news, weather, and sports databases as well.

DELPHI   Covers home banking, computer periodicals, news headlines, encyclopedias, electronic messaging, business and finance, and library and computer clubs.

DIALOG   This largest supplier of word-oriented databases is a division of Lockheed Missiles and Space Division. Most references are bibliographic. DIALOG has more than 100 million items of information stored. Magazine and journal articles can be searched and ordered on-line; document delivery takes from a few days to a couple of weeks. More than 320 databases are available, including chemistry, computer technology, science and technology, medicine, and biosciences. They reference more than 60,000 paper publications, and index more than 700.

Dow Jones   Owned by Dow Jones Inc., and has exclusive access to the full text of the *Wall Street Journal*. Databases include financial markets, financial information, and the *Academic American Encyclopedia*.

GEnie   A product of General Electric Information Services and CompuServe's closest competitor, this user-friendly service is growing rapidly and is now second to CompuServe in the number of subscribers. GEnie has databases in computing and high-technology news, educational resources, financial data, and product databases. GEnie also has a low-cost, after-hours service.

Knowledge Index   This service is a subsidiary of DIALOG and is an after-hours, discount service. Most of Knowledge Index's databases are bibliographic. There are 75 different databases in 13 categories that cover chemistry, computers, electronics, medicine, drugs, science, and technology. Best for beginners.

Mead Data Central   A division of Mead Corporation. Mead's original service was Lexis, a full-text database for lawyers. They also have Nexis, a full-text news service and Mead handles the *New York Times.* Lexpat, another of Mead's databases, has indexed more than 750,000 patents since 1975. Medis-Medical science is another of Mead's databases.

NEWSNET   NEWSNET carries about 300 newsletters grouped into 40 different industry groups, wide-ranging subject areas, and 10 leading wire services. Over 70 percent of NEWSNET's coverage is exclusive. It covers computers, management, metals and mining, research and development, telecommunications, chemistry, and energy. Many databases are in the field of computers.

ORBIT SDC/ORBIT   Owned by Burroughs, this service is bibliographic, with over 100 databases and more than 75 million citations. Specialties include patents, materials, science, chemistry, energy and the environment, engineering, and electronics.

VIDEOLOG   A subsidiary of Schweber Electronics, VIDEOLOG is basically an electronic catalog of components, manufacturers, and products. You can search for more than 750,000 components for both military and commercial applications. Components are from 700 manufacturers and 14,000 suppliers of products in more than 2,000 categories. A price quote can be obtained and the product can be ordered on-line, making a paper transfer unnecessary. A paper copy will be provided for your records if needed. Good graphics permit a display of response curves, outline dimensions, etc. VIDEOLOG also has electronic industry and new-product release news.

VU/TEXT   Provides full-text versions of 40 regional newspapers (36 are exclusive), plus selected articles from more than 180 regional business journals and newspapers. VU/TEXT also has the 20-volume *Academic American Encyclopedia* with more than 32,000 articles.

Wilsonline   H.W. Wilson produces printed indexes for libraries and has about 20 bibliographic databases, including the *Readers' Guide and Business Periodicals Index.*

Yellow Pages   Dun and Bradstreet's Electronic Yellow Pages is an electronic collection of more than 5,000 yellow pages from cities throughout the United States. It lists 6½ million entries.

DIGITAL DIPLOMAS

As libraries are computerized, it's only natural that universities and colleges follow. Some of the pioneers in these "electronic universities" offer on-line digital diplomas to people who cannot attend regular classes on campus, but who can go to class in their home using a PC and modem.

These classes have no schedule, and student/faculty conferencing can be held at mutually convenient times from the comfort of your home. Classes can be taken for credits toward degrees, or they can be audited for professional advancement.

The American Open University of the New York Institute of Technology offers a B.S. in Business Administration, General Studies, or the Behavioral Sciences. And you don't even have to visit their campus; all of your classes can be taken in your home. There is a one-time admission and matriculation fee of $150 and a charge of $85 per credit hour, which is about half what a resident student of the New York Institute of Technology pays.

Another electronic university, Connect.ed, offers a series of courses to fulfill the requirements for a 39-credit degree program leading to an M.A. in Media Studies. Connect.ed courses carry graduate and undergraduate credit from the New York School for Social Research. Courses can be taken at any time, 24 hours a day, 7 days a week, from any place in the world equipped with a PC, modem, and telephone. Graduate tuition costs $948 per course (for 3-credit courses) plus a $60 registration fee. Undergraduate (general credit) and noncredit tuition is $888 per course, and the registration fee is $20. Tuition covers all necessary charges except phone-line costs.

Nova University of Fort Lauderdale, Florida, offers masters degrees in Information Systems and Electronic Education, and Doctors of Arts in Information Sciences and Information Systems. Students must show up at the Fort Lauderdale campus twice a year for one-week institutes. Annual tuition is $4,500, and they recommend that a student budget an additional $2,500 per year for miscellaneous expenses.

DEVELOPMENTS

Some of the hardware developments that are making these dramatic changes in information access possible are discussed here.

Magnetic floppy diskettes with an optical servo can increase data storage 10 times — to 25 megabytes — over a conventional floppy disk. And 20 MB disks are economical and have become common in personal computers. A 20 MB disk has enough storage to hold 150 books of 500 pages each.

But the inventive people who develop these products have many more hardware innovations coming up. A 256 MB magneto-optical storage disk now available can hold 100,000 pages of text. That's enough for a small home library of 400 books.

Another promising development is called WORM — Write Once, Read Many (times). This is a potentially low-cost digital paper that can be used in place of a floppy disk and store as much as 700 MB. When used as digital tape, a 12-inch reel can store 1,000 gigabytes — enough to fill one billion sheets of paper with an average access time of 28 seconds. The cost is projected to be as low as ½ cent per megabyte, or one-tenth the cost of paper tape. And since it costs so little, digital paper can be a throw-away item.

An estimated less than 2 percent of the information mankind has generated is coded and stored in any kind of computer format. We are only seeing a small part of what lies ahead in the profession of information retrieval. New technologies are going to change our lives in many new and unexpected ways.

EXERCISES

1. Which on-line services would you query to obtain more information for your article topic?

2. If you have access to an on-line service at work, your college, your public library, or your home, contact one of the low-cost services and research your topic.

10

Good Writing

The greatest merit of style, of course,
is to have words disappear into thoughts.

Nathaniel Hawthorne

By now many of you will have concluded that writing of things technical can be fun and that you'd like to learn more about this fascinating and rewarding aspect of your profession and improve your writing even more. This chapter will show you techniques to use that will add a higher polish to your writing and instill in you the confidence to tackle a huge article for a first-rate periodical, perhaps even to start working toward the ultimate goal of most professionals: a book. This chapter gives you a graduate course in writing of things technical. You'll learn what good writing is and how to achieve it.

WHAT IS GOOD WRITING?

Good writing is writing that effectively communicates what the writer has to say to the reader, with a minimum of effort on the reader's part. Simple. You efficiently communicate something to someone.

Good writing uses many of the techniques you've learned throughout your life, plus what you've learned in this book and will learn in this chapter. Good writing has many ingredients:

- Logically organize the material
- Use good grammar
- Check spelling
- Use short sentences and paragraphs (vary their lengths)
- Above all, don't use big words!

Good writing is "invisible." By that I mean good writing helps your reader easily understand what you've written, but does not make obvious what techniques you've used to attain this effect. Properly used, these techniques are totally transparent to your readers.

HOW WE READ

First let's take a look at the basic way people read. The more mental energy it takes to dig meaning out of words, sentences, and paragraphs, the less energy readers have left to do something with what they learn and the more likely they will quit in exhaustion.

When we read technical material, we use energy in three basic ways:

1. To understand the words.
2. To understand the way the words relate to each other to form pictures and thoughts.
3. To understand the technical concepts described by the words and the accompanying visuals.

When we read fiction and nonfiction, we use mental energy only for the first two. Technical writing, however, adds a third and more difficult aspect: comprehending the technical concepts.

If you use short, simple, concrete words, you make the first step easy. If you use a variety of short, well-constructed sentences, you make the second task effortless. That eliminates two-thirds of the effort of reading, so your readers can concentrate all of their energy on understanding the technical concepts.

How do you do this? With good writing.

HOW WE UNDERSTAND

The basic element of writing, the basic symbol that paints a picture in a reader's mind, is the word. But the 500 most-used English words have a total of 14,000 dictionary definitions — that's an average of 28 meanings per word. And words can form radically different shades of meaning in different people's minds. For example, *house* can form a picture ranging from a tiny, rundown shack, to a condo, to a 50-room mansion. So you can see the importance of using the proper word.

Groups of words (phrases and sentences) modify and clarify the intrinsic meaning of the words they contain. Words should be used to sharpen the focus of the picture they are to create. *Man* gives you an out-of-focus, fuzzy, indefinite picture. *Fat man* brings the picture a little more in focus. *Short, swarthy, grossly fat man* sharply focuses the image the words are to convey.

The more specific the word, the more concrete, sharper, and more in focus the picture.

WORD ORDER

It's not only the words you use, but also the word order that is important. Words in your sentences must be arranged so that they can mean only what is intended. In the English language, changing the word order changes the meaning:

Man bites snake.
Snake bites man.

The proper word order is needed to paint the exact picture you want.

Another important aspect of helping your readers understand is consecutiveness: things happen in a given order.

Add one liter of water to the solution, shake vigorously for 10 seconds, then add one gram of KC1.

Using the proper order makes an idea easy to follow.

Still, words are not substitutes for thoughts. You must group words together to form a sentence. But no sentence, regardless of the words you use, is any better than the thought behind it.

A sentence should carry only a single meaning. The basic pattern of a sentence is:

Subject   *Verb*   *Object*

Thing - - - ->Doings - - - ->Result

Girl   - - - ->Throws - - - ->Basketball

Using the proper words in this basic SVO order makes your writing easier to understand because it's the sequence in which things happen. First you see the girl, then the throwing motion, then the basketball, the object being thrown.

If you say "The basketball was thrown by the girl," it's a backward construction. It jerks your mind out of gear. It's not the order in which the event occurred. First your mind focuses on *the basketball,* then on the passive *was thrown,* and finally on *the girl,* the most important part of the sentence. You have to unscramble it in your mind and put it back in the proper order before you can understand it. And this wastes your mental energy.

To make your sentence give a more complete picture, add a few more descriptive words: "The tall, blond girl lofted the basketball into the net."

You must arrange words in the proper grammatical form to carry the right meaning. An estimated 80 percent of poor writing stems from poor grammar.

In a study of 20 top writers (10 fiction and 10 nonfiction), over 75 percent of all the sentences in their writing used a subject-verb-object order: "Man bites dog." So, to be understood, it's best to use the common SVO order in most of your sentences.

We read in terms of words and we understand words within the limits of our vocabulary. Unless the message is contained within the area of the common vocabulary shared by the writer and the reader, as shown in FIG. 10-1, there is no communication. In addition, the words in the message must mean the same to both writer and reader.

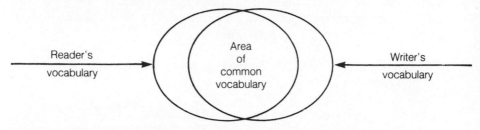

Fig. 10-1. Writer's/reader's vocabularies.

You don't need a large vocabulary for effective communication. An estimated 1,000 words cover about 85 percent of a writer's requirements on all ordinary subjects. Add some of the special words of your profession and you can certainly do most of your writing without having to resort to long, abstract, indefinite words. If you use as many of the 1,000 basic words as you can in place of complex words or jargon, your writing will be that much easier to understand.

GRAMMAR

You must use proper grammar to effectively communicate something to someone, not only to avoid misunderstanding, but also because your reader will notice improper grammar. Grammatical errors distract your reader's attention from the primary purpose of your writing: communicating technical information.

It's no grave sin to misuse a few *whos* or *whoms* because few people understand, or even care about, these finer points of grammar. But if you make obvious mistakes — such as using the wrong tense of verbs, incorrect order of words, or other serious basic errors in grammatical construction — your reader's mind stops understanding the intent of your article and starts looking for more errors in grammar. This distracts the reader's concentration on the technical content.

One of the best general guidelines for technical writing is to write as if you're taking your reader into your confidence. Try to create meaningful comparisons with things your readers are already familiar with. We learn best by analogy, by comparison, by contrast with things we are already familiar with.

SPELLING

When your document is sprinkled with misspelled words, your reader's mind is again diverted from understanding the technical content. Also, when people read a document full of typos, errors, and misspelled words, they begin to distrust the accuracy of the technical content, the visuals — everything associated with the document. With the variety of modern electronic dictionaries and with the excellent spelling programs now available in most word processors, spelling errors should soon become a thing of the past.

Of the approximately 20,000 commonly used words, only 1 percent are consistently misspelled. And of these 200, the 50 that cause the most trouble are listed, left to right, in order of difficulty:

grammar	argument	surprise
achieve	anoint	definitely
separate	desirable	development
existence	pronunciation	occasion
assistant	repetition	privilege
dependent	irresistible	consensus
accommodate	occurrence	conscience
commitment	embarrass	allotted
indispensable	liaison	proceed
harass	perseverance	ecstasy
antiquated	insistent	exhilarate
vacuum	ridiculous	nickel
oscillate	tyrannous	drunkenness
dissension	connoisseur	sacrilegious
battalion	prerogative	iridescent
inadvertent	genealogy	vilify
inoculate	dilettante	

If you can study and be extra careful with these words, you'll have a good start.

Concrete Words Are Best

Concrete words paint a more understandable picture than abstract words. But how do you identify concrete words? To test for concreteness, think whether you can:

- Feel it
- See it
- Hear it
- Taste it
- Smell it

Most importantly of all, does the word invoke one, and only one, clear picture in your and the reader's minds?

Set your words in concrete and you'll have a solid foundation for your article.

SHORT IS USUALLY BETTER

Short sentences and paragraphs have been proven to be effective ways to communicate ideas. A reader tries to comprehend an entire sentence in one gulp, looking ahead to the period (the rest stop) before pausing.

Avoid long, convoluted sentences. By the time your readers get to the end of such a sentence, they have forgotten what was at the beginning. Put the main message at the front of the sentence. If your statement has to be qualified, do that in the next sentence.

Paragraphs

A paragraph is a sentence or group of sentences that express and develop one major idea. A paragraph groups together sentences that confirm the same topic and combine to form a thought-unit. The end of a paragraph provides a longer pause for the reader to catch his or her breath. The reader's mind is geared to reading and understanding a single idea, feeling that once a paragraph is read, the idea has been completely presented and he or she deserves a momentary pause before tackling the next idea or paragraph.

Use paragraphs of varied length. You can use this technique to highlight the more important ideas with longer paragraphs, contrasting them with shorter paragraphs that convey less important ideas.

Experience has shown that readers of technical material can grasp material most readily when it is presented in units of 75 to 200 words. With an average of 10 words per typed line, that's a range of 7.5 to 20 lines to allow per paragraph. Again, don't take this guideline as gospel and insist that all paragraphs fit between these limits. These numbers are average numbers, a method to check your writing along the way. Use some one-sentence paragraphs and some much longer ones. Vary their lengths for variety.

The most effective way to introduce the idea that the paragraph is to cover is to use a topic sentence as the first sentence in the paragraph. The topic sentence should contain these essential elements:

- The writer's theme or idea
- IIis or hcr viewpoint
- The topic of the paragraph

For example, consider this topic sentence: "Children's television cartoon programs are harmful." The topic sentence conveys the theme that children's television cartoons will be discussed. The viewpoint is subjective, the author's opinion. The topic is the harm that television cartoons can do to children.

A topic sentence indicates the direction an idea is to take. It helps guide the reader to the central purpose of the paragraph. It leads the reader through the development of a topic so that he or she reaches the same conclusion the writer intended. The topic sentence isn't always placed at the beginning; it can be at the end if the paragraph is developed by induction. If a writer considers a point vital, he or she might emphasize the idea at both the beginning and the end of a paragraph. Wherever it is placed, the topic sentence should dominate the paragraph, and the rest of the paragraph should serve mainly to develop it.

The topic sentence is often followed by one or more subtopic sentences that further develop, clarify, and prove the thesis of the topic sentence. Facts, statistics, or quotes (if experts can be included as part of the proof) form the subtopics. Without the proof, the reader will wonder what is the basis for the idea. The last sentence in the paragraph can be a conclusion, a call to action, or a transitional phrase to link to the next topic or idea.

Transitions

Transitions are the bridges that link paragraphs. They provide continuity so the reader can move from one topic to another without feeling a jolt in continuity. Transitions connect what has been said with what is about to be said.

Some more commonly used transitions are:

- Time   *Later, the next time, finally, next, concurrently, after.*
- Place   *In the next room, farther down the hall, at our plant in Texas.*
- Piling up of detail   *And, also, furthermore, in addition, moreover, besides.*
- Contrast   *But, however, though, although, nonetheless, yet.*
- Illustration   *For example, to illustrate, in particular, for instance.*
- Cause/Effect   *Thus, therefore, in conclusion, as a consequence, as a result, consequently.*
- Comparison   *In a similar way, likewise, similarly, here again, consequently.*
- Concession   *Although, even though, since, though.*
- Summary   *To sum up, in brief, in short.*
- Repetition   *In other words, that is, as has been stated.*

- Numbered or lettered steps   *Used in a procedure.*
- Repetition of and references to preceding ideas, key words, or phrases.

PUNCTUATION

Punctuation is the written equivalent of the rhythm of our speech. It takes the place of the rise and fall of our voice, and the pauses and emphasis in our speech. The major function of punctuation is to make writing clearer and easier to read. Punctuation helps express, question, emphasize, surprise, and conclude.

When we speak, our listeners learn as much from the way we speak as they learn from the words we speak. We gesture, screw up our face, raise or lower our voice, pause, speak fast or slow, speak high or low, use our hands, raise our eyebrows, and use our facial expressions to add emphasis to words.

Comma   About half of the total number of punctuation marks used in writing are commas. We use commas to separate words or phrases to avoid confusion and ambiguity. It's the main device by which a grouping of words, phrases, and clauses is indicated. The comma gives a short pause to illustrate and emphasize this separation.

Period   The period signifies the end of a sentence. It is equivalent to a longer pause by a speaker. The period used in abbreviations and acronyms is rapidly disappearing from printed matter because it confuses readers. Readers mistakenly conclude that they've reached the end of the sentence. It's better to spell out the words, rather than use abbreviations, except for some of the more commonly used ones, such as *etc.* You can also use the abbreviation or acronym without the periods, as in *USA.*

In spite of the disappearance of periods in abbreviations, technical writing experts agree that the period is not used enough in technical writing. They believe it should be used more often to end sentences faster. Sentences should be shorter.

Semicolon   The semicolon is sort of a half-period. It gives a pause of a length between the comma and the period. The semicolon is optional; you don't have to use it. In fact, most great writers seldom, if ever, use it. Some writers have gone an entire lifetime without getting acquainted with the semicolon.

Colon   The colon introduces a list of details or an explanation. It is equivalent to a gesture or a lift of the eyebrow in a speaker.

Exclamation Point   The exclamation point is the equivalent of a speaker emphasizing a point with a raised voice or a gesture.

Question Mark   The question mark takes the place of the voice being raised at the end of a question.

In spite of all of the confusing and sometimes vague rules about punctuation, the best guideline for proper punctuation is common sense. Forget what the grammarians say about periods, commas, and quotes. If you feel that putting a stop in a specific place in your sentence will clarify it for your reader, do so. If you find you're using too many commas in your sentences, perhaps your sentences are too long and confusing. Shorten or reshape your sentences to make them easier to understand.

Like all tools, the use of punctuation marks can be overdone. Too many stops produce a jerky style, hindering your reader from understanding your writing.

THE CURSE OF BIG WORDS

Bertrand Russell stated in his *How I Write,* "Big men write little words, little men write big words."

Technical ideas are difficult enough to understand without a writer complicating them even more by trying to show off his or her vocabulary. You should write to *express,* not to *impress.* Even the most complex scientific ideas can be presented by simple (three-syllable or less) words. Some of the most famous and brilliant scientists of all time, such as Charles Darwin, Louis Pasteur, and Marie Curie, explained their complex concepts in words so simple that the public could understand and appreciate the significance of their discoveries. Undoubtedly their clear writing contributed as much to their fame as did their discoveries.

It takes a little mind (with a dictionary and a thesaurus) to write with big words. It takes a big mind to express complex thoughts and ideas in short words.

DON'T OVERDO IT

If overdone and abused, these techniques lose their effectiveness. If you use only one or two sentences in every paragraph throughout your document, or ten-word "Dick and Jane and Spot" sentences, you again distract your reader from understanding the technical message you want to convey. Your task is to use the techniques of good writing to communicate smoothly, without your reader becoming aware of the techniques you're using. The techniques of good writing should be invisible. Your reader shouldn't know what you're doing to him or her, nor how you're doing it. They should just be able to enjoy your writing, and learn from and understand what you've written.

RETENTION CURVE

One extremely important aspect of the manner in which a person's mind works is illustrated in FIG. 10-2. This curve is valid for all people, of all cultures, regardless of education. It's a fundamental fact of human nature.

Stated simply, the curve illustrates that The mind remembers best what it experiences, hears, and reads first and last. This fascinating fact holds true for

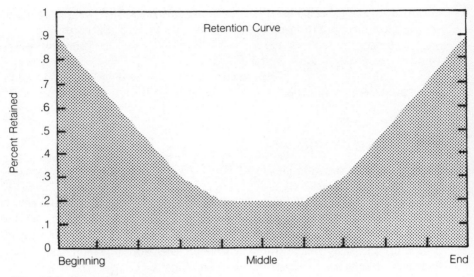

Fig. 10-2. Retention curve.

a book, a movie, a speech, a poem, an article, your life's experiences . . . whatever you see, hear, or read.

That's why the HEY!, YOU!, and SO! sections of your document are so important. Your reader will best remember the beginning and ending sections, so the summary and conclusion of your article are worth your best efforts.

The same phenomenon holds true for both sentences and paragraphs. The first and last parts of sentences and paragraphs are best remembered by your readers. That's why the topic sentence (the first sentence in your paragraph, which introduces the paragraph's contents) is so important. And the final sentence of your paragraph should be a sort of conclusion. Your readers should have the feeling of being introduced to a concept in the first sentence and should end the paragraph by understanding that concept.

This important aspect of the human mind is used in many ways, even by astute politicians. It is an established fact that on election ballots, when a person's name is in the middle of two opponents' names, that person will lose several percentage points off his or her total vote. That is true, regardless of the candidate's qualifications. People simply vote more for the names at the top and bottom of a listing.

You can take advantage of the retention curve in many ways. Keep it in mind when you want to emphasize one or more points. Put the most important points in the first part of your writing, in your first paragraphs, in the first sentences of your paragraphs, and at the beginnings and endings of your sentences. And save some of your best writing for the conclusion.

You can also use the phenomenon of the retention curve to conceal information, bad data, meager results, etc. Bury bad information, which you must include for completeness, in the middle of the document. A strong opening paragraph often will negate the bad information in the middle.

RULES OF THUMB

There are a number of rules of thumb that you'll find useful in writing to help you figure the number of words, the number of pages, etc. It takes about two pages of handwritten material to equal about one page of typed, double-spaced text. This can vary a lot depending on a person's handwriting style. If you write your material long-hand, you should experiment and make a note of its typed equivalent for future reference to properly gauge your writing length. Normal handwriting, double-spaced, includes about 125 words per page. A normal, double-spaced, typewritten page contains about 250 words, that is, about 25 lines of 10 words per line.

When you write an article for a technical periodical, add up the number of double-spaced, typewritten pages, photos, drawings, graphs, tables, etc. that you have generated. Then divide by three to get the number of periodical pages (technical periodicals usually pay by the number of pages).

Also, an article should have about one visual per finished page (or for each two or three typewritten pages) of text so the article will have some white space around it to make the layout more attractive.

Clichés to Avoid

Avoid clichés in writing of things technical. I'm including some of the most often abused ones here:

What the writer said	*What the writer should have said*
The purpose of this report is . . .	This report describes . . .
This report is submitted to . . .	This report sums up . . .
In order to conduct . . .	To conduct . . .
(*In order* can be left out of *all* writing.)	
It is recommended that the formula should be . . .	The formula should be . . .
(Watch out for *It . . . that* constructions; they waste a lot of words).	
It may be said that copper is . . .	Copper is . . .
It is interesting to note that the welded corners . . .	Welded corners will . . .
In accordance with . . .	Per . . .
As defined in . . .	Per . . .
Shall have the capability of converting . . .	Shall convert . . .

The generator shall be tested per . . .	Test the generator per . . .
Due to the fact that . . .	Since . . .
Demonstrates that there is . . .	Shows . . .
During such time . . .	While . . .
If the developments are such that . . .	If . . .
In consideration of the fact that . . .	Since . . ./Because . . .
Make as approximation as to how . . .	estimate . . .
Reduced to basic essentials . . .	simplified . . .
The purpose of this report is . . .	This report shows . . .
Would seem to suggest . . .	Suggests . . .
For the purpose of . . .	For . . . to . . .
Is designed to be . . .	Is . . .
In close proximity . . .	Close to; near . . .
Subsequent to . . .	After . . .
In the event that . . .	If . . .
In order to . . .	To . . .
Involves the use of . . .	Employs . . . uses . . .

IN CONCLUSION

In engineering and science, feedback helps stabilize a device. In our personal relationships it also helps stabilize and develop rapport among people. In writing, an author needs feedback from the readers to make future editions of a book better.

I'd like to hear your comments on how you liked the book, what you did and did not like, what I communicated well, and what I failed to communicate properly. Just write to me, in care of the publisher, and I'll use your feedback to improve future editions of this book.

Thank you. I hope to see all of you in print, soon.

EXERCISES

1. Calculate the fog index of your article (See chapter 2).

2. Calculate the fog index of a *Reader's Digest* or *Wall Street Journal* article.

3. Calculate the fog index of a leading scientific journal in your profession.

Appendix A
Reference Sources

The following categories of reference books are listed alphabetically in this appendix

- Abstracts and indexes
- Almanacs (yearbooks)
- Biographical dictionaries
- Books
- Companies
- Dictionaries, general
- Dictionaries, scientific
- Encyclopedias, general
- Encyclopedias, scientific
- Government publications
- Handbooks
- People
- Periodicals

For the reference books that follow, the title is listed first, followed by the author (if one is listed), then the publisher. A date such as (1941–) means the book was first published in 1941 and is still being published.

ABSTRACTS AND INDEXES

In this section you'll find the most up to date periodical, report, and document references. All of these abstracts and indexes, plus many more, are available on the various on-line systems.

Abstracts and Indexes in Science and Technology, 2nd ed. Metuchen, NJ: Dolores B. Owen, Scarecrow Press, 1985. Gives descriptions of 223 abstracts and indexes arranged in 11 subject categories, including on-line databases.

Applied Science and Technology Index, Bronx, NY: H. W. Wilson. (1958-). Published monthly, except August, with quarterly and annual accumulations. From 1913 to 1957, this index was known as the *Industrial Arts Index.* Arranged alphabetically by subject only. Lists articles from over 300 English-language periodicals in applied science and technology. Excellent source for the latest articles on new and old science and technology.

Astronomy and Astrophysics Abstracts, New York: Springer-Verlag. (1969-) Semiannual. Subject and author index to periodicals throughout the world printed in English, with some in French and German.

Biological and Agricultural Index, Bronx, NY: H. W. Wilson. (1964-) Alphabetical subject index to approximately 200 English-language periodicals in biological and agricultural sciences. Also has book reviews.

Biological Abstracts, Philadelphia, PA: BIOSIS. (1926-). Worldwide reporting of research in life sciences. Principal abstracting journal for biology. Contains abstracts, author index, biosystematic index, generic index, and subject index.

Business Periodicals Index, Bronx, NY: H. W. Wilson (1958-). Indexes by subject, about 275 English-language periodicals in all fields of business. Index is issued monthly and accumulated into annual volumes.

Chemical Abstracts, New York: American Chemical Society (1907-). Subject and author index to abstracts from more than 14,000 periodicals in more than 50 languages. Known as the "key to the world's chemical literature," *Chemical Abstracts* is the principal source for titles and abbreviations in the physical sciences, life sciences, and engineering.

Chemical Titles, Columbus, OH: Chemical Abstract Service (1960-). Author and keyword computer-produced index to 700 periodicals covering pure and applied chemistry and chemical engineering.

Computer and Control Abstracts: Science Abstracts, Series C, London: Institute of Electrical Engineers, and New York: Institute of Electrical and Electronic Engineers. Monthly indexes and abstracts international electrotechnology literature by author and subject. Separate index to bibliographies, books, reports, conference proceedings, and patents.

Computer Abstracts, Technical Information Co. (1957). Offers abstracts of books, periodical articles, conference proceedings, U.S. Government reports, and patents in classified arrangements. Annual and subject indexes.

Engineering Index, New York: Engineering Information Inc. (1884-). Monthly with annual accumulations. The basic English-language abstracting service. Abstracts over 2,700 professional and technical journals, as well as reports and proceedings published in 20 or more languages. A general index to engineering literature, arranged by subject, with an author index.

General Science Index (GSI), Bronx, NY: H. W. Wilson (1978–). Indexes more than 100 general science periodicals not completely covered by other abstracts and indexes in astronomy, chemistry, electricity, mathematics, physics, etc.

Index Medicus, Washington, DC: Washington National Library of Medicine (1960–). Contains subject, name, and bibliography of medical reviews. Over 2,000 periodicals are indexed, either completely or selectively.

International Aerospace Abstracts (IAA), Washington, DC: American Institute of Aeronautics and Astronautics. (1961–). Semi-monthly, with annual accumulations. Includes books, periodicals, conference papers, and translations. Indexed by subject and author. Companion service to *Scientific and Technical Aerospace Abstracts*, which covers "unpublished" material on the same topic.

Mathematical Reviews, Providence, RI: American Mathematical Society (1940–). Subject index to mathematical periodicals and books. Arranged by broad subject, with abstracts for most entries. Comprehensive coverage of the pure and applied mathematics literature.

Metals Abstracts, Metals Abstracts Trust (1968–). Author index to abstracts from 1,000 periodicals throughout the world, covering all aspects of the science and practice of metallurgy and related fields.

New York Times Index, New York: New York Times (1913–). Indexes all articles published in the *New York Times*, preserved on microfilm and carefully indexed by subject alphabetically. Semi-monthly.

Nuclear Science Abstracts — Energy Research Abstracts, Oak Ridge, TN: U.S. ERDA (1948–1976). Covers technical reports of ERDA and its contractors in nuclear science and technology. Has detailed author and subject index. Covers books, articles, and papers on nuclear science.

Physics Abstracts, Science Abstracts, Series A, Institute of Electrical Engineering, London; and Institute of Electrical and Electronic Engineering, New York (1898–). Monthly. Separate index to bibliographies, books, conferences, patents, and reports.

Reader's Guide to Periodical Literature, Bronx, NY: H. W. Wilson (1900–). Best-known popular periodical index. Author and subject index to general interest periodicals (limited, however — only covers about 174 of the thousands of magazines in circulation), including a few scientific periodicals published in the United States. From 1802–1906, it was titled *Poole's Index to Periodical Literature*.

Science Citation Index, Philadelphia, PA: Philadelphia Institute for Scientific Information (1961–). Lists the reference or coded author with his or her work for over 3,800 journals and monographic series. International. Provides access to books, papers, articles, and reports being cited in currently published papers.

Scientific and Technical Aerospace Abstracts (STAR), Washington, DC: U. S. Government Printing Office (1963–). Comprehensive abstracts worldwide of unpublished reports on the sciences and technology of space and aeronautics, especially NASA reports. Indexed by subject, report number, accession number, and individual and/or corporate authors.

Technical Abstracts Bulletin (TAB), Washington, D.C.: Defense Documentation Center (1953–). Abstracts and announces classified and unclassified or limited-distribution documents produced by the Department of Defense and its contractors and acquired by DDC.

U. S. Government Research and Development Reports, Washington, D.C.: Clearinghouse for Federal Scientific and Technical Information (1946–). Abstracting and announcement bulletin covering reports, including progress reports of R&D under government auspices.

ALMANACS

Statistics, summaries, general interest facts, updated every year. A surprising amount of information in each single volume.

Information Please Almanac, New York: Simon and Schuster (1947–). More legible and easier to use than the *World Almanac,* but its coverage is not as complete.

McGraw-Hill Yearbook of Science and Technology, New York: McGraw-Hill, annual. Reviews the past year's works in science and technology. Supplements the McGraw-Hill *Encyclopedia of Science and Technology.*

World Almanac and Book of Facts, New York: World Telegram (1868–). Annual. Often cited as the best-selling American reference work, as well as the most comprehensive almanac and most frequently useful.

BIOGRAPHICAL DICTIONARIES

Biographies are useful when you need to look up names and affiliations of experts in your field you want to consult with or write about.

American Men and Women of Science, New York: R. R. Bowker, 7 volumes (1982). Brief biographies of over 130,000 living U.S. and Canadian scientists, including positions held, education, area of specialty, etc.

Biography Index, Bronx, NY: H. W. Wilson, Quarterly (1947–). Name and profession index to biographical material in 1,500 periodicals and books. The most comprehensive index in the field. Arranged alphabetically by the name of the subject of the biography.

Chamber's Biographical Dictionary, New York: St. Martin's (1969). A good dictionary covering the great of all nations, both living and dead.

Dictionary of Scientific Biography, New York: Scribner (1970–80). 16 volumes. Comprehensive, covers all periods from antiquity to present and includes only deceased scientists. Entries give place and date of birth and death, and brief summaries of the individual's contributions to science.

Who's Who in America, Wilonette, IL: Marquis (1899–). Biennial. Best-known and most useful general dictionary of contemporary biography. People included from all fields, science, education, business, etc.

World Who's Who in Technology Today, J. Dick (1984). 5v. Covers electronics and computer science, physics and optics, chemistry and biotechnology, mechanical and civil engineering, energy and earth sciences.

BOOKS

If the information you're searching for is in a book, a number of potential sources exist for locating that book, in addition to using an electronic or manual card catalog.

Book Review Digest, Bronx, NY: H. W. Wilson (1905–). Gives digests of book reviews taken from some 75 American and English general interest magazines. Arranged by title, it carries title and subject indexes.

Book Review Index, Detroit, MI: Gale Research Co. (1965–). Author index to reviews of books in more than 450 general interest magazines.

Books in Print, New York: R. R. Bowker (1948–). Annual. Lists the books from some 3,600 publishers that are in print and can be purchased. Has author, title, and subject listings in separate volumes.

A Brief Guide to Sources of Scientific and Technical Information, Saul Herner, Arlington, VA: Information Resources Press. Guide to major sources of technical information, intended for the engineer and scientist. Indexed.

Cumulative Book Index, Bronx, NY: H. W. Wilson (1898–). Lists by subject and author English-language books in print of general interest, including the sciences. Arranged by author, title, and subject.

New Technical Books, New York: New York Public Library (1915–). Classed subject arrangement includes table of contents and annotations for each book. Subject emphasis is on the pure and applied sciences, mathematics, engineering, industrial technology, and related disciplines.

Technical Book Review Index, Washington, DC: Special Libraries Association (1935–). Monthly except July and August. Index to book reviews appearing in scientific, technical, and trade journals. Provides brief quotations from reviews. Best book review index for scientific books. Arranged by author of book reviewed.

COMPANIES

Sometimes you might need to contact specific companies in your field, or in a related field, and need their names and addresses. Try the excellent directories listed here.

Standard and Poor's Register of Corporations, Executives, and Industries, New York: Standard and Poor's (1928–). 3v. Listings of over 45,000 companies. This is the "who's who" of American companies. One volume is an alphabetical listing by company name; a second volume is a directory of executives and board members; the third volume is a set of indexes to the first volume.

Thomas' Register of American Manufacturers, New York: Thomas Publishing Co. (1906–). Annual. Massive listing. Provides a comprehensive, detailed guide to the full range of products manufactured in the United States. Lists manufacturers' names, addresses, products, and trade names.

The Yellow Pages of Your Telephone Directory. Not only for your immediate city, but telephone directories of many other cities are kept in your local library.

DICTIONARIES, GENERAL

For general definitions (good for crossword puzzles, too), these are the best:

- *A Dictionary of American English*
- *Funk and Wagnalls*
- *Oxford English Dictionary.* A classic! *Warning:* you can easily get captivated when using this fascinating dictionary. It is the most authoritative dictionary of the English language and gives the *etymology,* or history, of the words.
- *The Random House Dictionary of the English Language*
- *Roget's International Thesaurus*
- *Webster's 3rd New International Dictionary,* Merriam-Webster Inc. 2,600 pages. Can be used for years.

DICTIONARIES, SCIENTIFIC

For quick definitions of terms that might puzzle you as you search through other literature, check these.

Aviation/Space Dictionary, The, Larry Reithmaier, Fall River, MA: Aero (1989). Definitions related to aircraft armament, power plants, airline operations, and air traffic control. A good, basic aerospace reference work.

Basic Dictionary of Science, New York: Macmillan (1966). A dictionary of 25,000 terms, followed by lists of abbreviations used in science. Chemical elements with their atomic weights and numbers. Written for those with little or no scientific background.

Concise Chemical and Technical Dictionary, New York: Chemical Publishing (1974). About 90,000 entries covering trademark products, chemicals, drugs, and terms.

Dictionary of Computers, Data Processing, and Telecommunications, New York: John Wiley (1984). Definitions of some 10,000 words and phrases. Has appendix of Spanish and French equivalents.

Thesaurus of Engineering and Scientific Terms, Engineers Joint Council (1967). 690 pp. Has over 20,000 entries of engineering and related scientific terms and their relationships.

IEEE Standard Dictionary of Electrical and Electronic Terms, Institute of Elect. and Electronic Engr. New York: John Wiley and Sons, Inc. Excellent dictionary, an official source of definitions taken from standards of IEEE, ANSI, and IEC.

McGraw-Hill Dictionary of Scientific and Technical Terms, New York: McGraw-Hill (1984). 1781 pp. Over 98,500 terms with 115,000 definitions from technology and science are defined clearly and concisely.

Modern Dictionary of Electronics, New York: McGraw-Hill (1984). Brief descriptions of several thousand terms in current use. Has pronunciation guide and defines widely used symbols.

ENCYCLOPEDIAS, GENERAL INTEREST

If you know little or nothing about your topic, a general encyclopedia is often a good place to start. Occasionally they have excellent summaries on what you're looking for and, more importantly, good references to other documents having greater depth. Usually they're well cross-referenced. Encyclopedias are usually more up to date than books because they're revised more often.

Academic American, Danbury, CT: Grolier, Inc. 21 v. Falls somewhere between the *World Book* and the *Britannica* in scope and depth of treatment. The full text of the encyclopedia is available on-line through a number of commercial vendors. Intended for students in junior high, high school, and college, as well as inquisitive adults.

Colliers Encyclopedia, New York: Macmillan. Emphasis on simple explanations. Strong in contemporary science. Third in size to *Americana* and *Britannica,* but the most current, best indexed, and easiest to read of the three. Style is popular, clear, and concise.

Consise Columbia Encyclopedia, The, New York: Columbia University Press. One volume, capsule size for quick reference. Contains more separate entries than most English-language encyclopedias.

Encyclopedia Americana, Danbury, CT: Grolier, Inc. 30v. Full and scholarly. Strong on American topics, especially strong in science and technology. Good, comprehensive encyclopedia for general use.

Encyclopaedia Britannica, Chicago, IL: Encyclopedia Britannica. Full and scholarly. Good coverage of both British and American topics. The most famous encyclopedia in English. Long, detailed articles on many and diverse subjects.

Funk and Wagnalls New Encyclopedia, New York, Funk and Wagnalls. 27 v. Serves general family needs. Provides brief background on a wide variety of topics and is written in a clear, popular style. Useful, inexpensive choice.

World Book Encyclopedia, Raleigh, NC: Fields Enterprises Corp. Ranges from young people to general adult in content. Keyed to school curricula.

ENCYCLOPEDIAS, SCIENTIFIC

Scientific encyclopedias are good for general, basic information about a wide variety of subjects. However, they are not detailed technically.

Concise Encyclopedia of the Sciences, New York: Facts on File (1978). Complete guide to the language and history of science and technology. Covers the most commonly used words of science and technology, with background material.

Encyclopedia of the Biological Sciences, New York: Van Nostrand Rheinhold (1970). 1027 pp. Articles on instruments in theory and practice, techniques of microscopy, and preparation disciplines. List of references follows each article.

Encyclopedia of Chemical Technology, New York: John Wiley (1980–84) 24 v. Articles written by specialists, includes bibliographies. Uses SI as well as English units.

Encyclopedia of Computer Science and Engineering, New York: Van Nostrand Rheinhold Co. (1983). 1,664 pages of descriptions of basic computer terms. 550 articles for the nonspecialist. Cross-referenced; excellent index.

Encyclopaedic Dictionary of Physics, Elmsford, NY: Pergammon Press (1961–1975). 9 v. Scholarly work, alphabetically arranged. Articles on general, nuclear, solid-state, molecular, chemical, metal, and vacuum physics.

Encyclopedia of Engineering Signs and Symbols, New York: Odyssey Press (1965). A comprehensive dictionary of signs, symbols, and abbreviations used in engineering.

Harper Encyclopedia of Science, New York: Harper and Row (1967). 1379 pp. One volume. Excellent, readable, covering a wide range of topics. Biographies are included. Entries are brief but informative.

Encyclopedia of Polymer Science and Technology, New York: John Wiley Interscience (1964–72). 14 v. Contains equations, graphs, tabular data, and extensive bibliographic references. Covers all aspects of polymer science (physics, chemistry, biology) and engineering.

McGraw-Hill Encyclopedia of Science and Technology, New York: McGraw-Hill (1982). 15 v. Covers all major scientific subjects. Not too technical, easy to read. Some 7,600 articles by 2,500 scientists and engineers. Many illustrations. Alphabetically arranged with cross references. This is the best of the science and technology encyclopedias, kept up to date by the *McGraw-Hill Yearbook of Science and Technology.*

Van Nostrand Scientific Encyclopedia, New York: Van Nostrand Reinhold (1988). 3264 pp. Defines about 16,500 terms in the physical sciences, computer technology, electrical engineering, electronics, pure and applied math, etc. Over 2,000 illustrations, arranged alphabetically, with excellent cross references.

GOVERNMENT PUBLICATIONS

Guide to Popular U. S. Government Publications, Englewood, CO: Libraries Unlimited (1986). 432 pp. References 2,900 titles of government documents, most of which have been published since 1978, with a brief description of each document's contents.

Monthly Catalog of U. S. Government Publications, Washington, DC: U. S. Govt. Printing Office (1895–). Comprehensive index. Lists all publications of all government departments, bureaucracies, and agencies. Published monthly, with an annual cumulative index. Gives author, title, publication data, price, and availability. Does not include classified documents.

Subject Guide to U. S. Government Reference Sources, Chicago, IL: American Library Association (1985). Subject index, plus a directory. Annotated bibliography of key government documents on general topics, including science and technology. Over 1,300 entries.

U.S. Government Reports Announcement and Index, Springfield, VA: U. S. Dept. of Commerce, National Technical Information Service (NTIS) (1971–). Semimonthly, with annual accumulations. Indexes and abstracts unclassified reports of U. S. government contractors in the public and private sector. Indexed by individual and corporate author, subject, report number, and accession number.

HANDBOOKS

Computer Dictionary and Handbook, Indianapolis, IN: Howard Sams (1980). Dictionary and handbook of 22,000 terms in electronic data processing, information technology, computer sciences, and automation.

CRC Handbook of Chemistry and Physics, Boca Raton, FL: CRC Press (1913–). An indispensable tool for chemists and physicists, this is a ready reference book of chemical and physical data.

Handbook of Mathematical Tables and Formulas, Richard S. Burington. New York: McGraw-Hill (1973). 500 pp. Quick reference to many vital tables of data in math, engineering, physics, chemistry, and physics.

PEOPLE

Authors of articles   Why not contact the author of one or more of the reports, books, etc. in your field of interest that you found particularly informative. He or she will usually cooperate and bring you up to date on what has been happening since the article was published.

Magazine editors   They're an excellent source for information about what's going on in a certain field, and they can direct you to experts in your specialty.

People at other companies   Often even your competitors will consult with you on a problem, particularly if it's on a person-to-person basis, rather than on a company-to-company basis.

Your friendly librarian   Librarians are always willing to help and can locate even the most obscure facts, or anything you need, in books you never even dreamed existed.

Your fellow employees   Check around your company; you might find an expert or two.

PERIODICALS

When you need to search specific periodicals in your field of interest, the following books will help you locate them.

Catalog of Scientific and Technical Periodicals (2nd ed. 1965), Washington, DC: Smithsonian Institute (1665–1895). Titles of more than 8,600 pure and applied science periodicals published throughout the world, from the rise of literature to the present time.

Business Periodicals Index, Bronx, NY: H. W. Wilson (1958–). Monthly except July. Subject index to about 275 English-language periodicals in all areas of business, computers, accounting, marketing, communications, etc. Arranged alphabetically by subject.

Encyclopedia of Associations, Detroit, MI: Gale Research Company (1956–). 5 v. Excellent reference. Lists some 18,000 national and international organizations. Includes scientific, engineering, and technical associations. Has key word, geographical, and executive index, research activities, and funding programs.

Gale Directory of Publications (Formerly Ayer Directory of Publications), Detroit, MI: Gale Research, Inc. (1989). Index of 25,000 newspapers and magazines published in the United States, Canada, and a few other countries.

Scientific, Engineering, and Medical Societies Publications in Print, New York: R. R. Bowker (1980–81). 626 pp. Lists publications published by 301 technical societies. Arranged alphabetically by society. Author index and periodical title index.

Standard Periodical Directory, New York: Oxbridge (1964/64–). The largest authoritative guide to U. S. and Canadian publications. Information on more than 65,000 publications. Alphabetical arrangement, with index of titles and subjects.

Ulrich's International Periodical Directory, New York: R. R. Bowker (1932–). 2 v. Lists over 70,000 currently published periodicals from more than 120 countries. Grouped by subject, with title and subject indexes. The most comprehensive of scientific (especially scholarly) publications from around the world. Also cites the indexing or abstracting services for each periodical. A very valuable reference.

Appendix B
List of Periodicals

Most of these periodicals have writer's guidelines. To request a copy, include a cover letter and an SASE.

68 Micro Journal
Computer Publishing, Inc.
5900 Cassandra Smith Rd.
Hixson, TN 37343–0794

73 Amateur Radio
70 Route 202, North
Peterborough, NH 03458–1194

A+
Ziff-Davis Pubs. Co.
950 Tower Lane
Foster City, CA 94404

AOPA
Aircraft Owners and Pilots Association
Pilot Aircraft Owners
421 Aviation Way
Frederick, MD 21701

Adhesives Age
6255 Barfield Rd.
Atlanta, GA 30328

Aerospace America
American Institute of Aeronautics and
Astronautics
370 L'Enfant Promenade, S.W.
Washington, D.C. 20024

Aerospace Engineer
400 Commonwealth Dr.
Warrendale, PA 15096

Ahoy
Haymarket Group Ltd.
45 W. 34th St., Suite 407
New York, NY 10001

*Air Conditioning, Heating &
Refrigeration News*
Business News Publishing Co.
P.O. Box 2600
Troy, MI 48007

*Air & Waste Management
Association*
P.O. Box 2861
Pittsburgh, PA 15230

Airport Services Management
50 S. Ninth St.
Minneapolis, MN 55402

American Biotechnology Lab
International Scientific
Communications, Inc.
30 Controls Dr.
P.O. Box 870
Shelton, CT 06484–0870

American Ceramic Society Bulletin
757 Brooksedge Plaza Dr.
Westerville, OH 43081–6136

*American Machinist Penton
Publications*
1100 Superior Ave.
Cleveland, OH 44114

AMERICAN MANAGEMENT ASSOCIATION
135 W. 50th St., 15th Floor
New York, NY 10020–1201

AMERICAN SCHOOL AND UNIVERSITY
401 N. Broad St.
Philadelphia, PA 19108

AMERICAN SOCIETY FOR ENGINEERING
EDUCATION
11 Dupont Circle, Suite 200
Washington, D.C. 20036

*American Tool, Die and Stamping
News*
Eagle Publications Inc.
31505 Grand River, No. 1
Farmington, MI 48024

America's Textile
2100 Powers Ferry Rd., Suite 125
Atlanta, GA 30357

Analytical Chemistry
American Chemical Society
1155 16th St., N.W.
Washington, DC 20036

Animal Kingdom
New York Zoological Society
185th St. and Southern Blvd.
Bronx, NY 10460

Appliance Manufacturer
Corcoran Communications, Inc.
29100 Aurora Rd.
Solon, OH 44139

Applied Optics
Optical Society of America
1816 Jefferson Place, N.W.
Washington, D.C. 20036

Applied Radiology
30 Vreeland Rd.
Florham Park, NJ 07932

Architectural & Engineering Systems
Softek Communications, Inc.
760 Whalers Way
Suite 100, Bldg. A
Fort Collins, CO 80525

Architectural Record
McGraw-Hill Information Systems
1221 Avenue of the Americas
New York, NY 10020

Architecture
American Institute of Architects
1735 New York Ave., N.W.
Washington, DC 20006

ASHRAE Journal
1791 Tullie Circle N.E.
Atlanta, GA 30329

The Asphalt Contractor
Specialty Publications Corp.
552 S. Brookside
Independence, MO 64053

Assembly Engineering
Hitchcock Publishing Co.
25 W. 550 Geneva Rd.
Wheaton, IL 60188

Astronomy
1027 N. Seventh St.
Milwaukee, WI 53233

Automatic Machining
100 Seneca Ave.
Rochester, NY 14621

Automation
Penton Publishing
1100 Superior Ave.
Cleveland, OH 44114

Automotive Engineering
Society of Automotive Engineers, Inc.
400 Commonwealth Dr.
Warrendale, PA 15096

Automotive Industries
Chilton Company
Chilton Way
Radnor, PA 19089

Avionics
Phillips Publishing, Inc.
7811 Montrose Rd.
Potomac, MD 20854

*Barron's National Business
and Financial Weekly*
200 Liberty St.
New York, NY 10281

Better Roads
P.O. Box 558
Park Ridge, IL 60068

Bio Science
730 11th St., N.W.
Washington, DC 20001–4584

Boxboard Containers
MacLean Hunter Publishing Co.
29 N. Wacker Dr.
Chicago, IL 60606

Broadcast Engineering
Intertec Publishing Co.
P.O. Box 12901
Overland Park, KS 66212

Broadcast Management/Engineering
295 Madison Ave.
New York, NY 10017

Bulletin of the Atomic Scientists
6042 South Kimbark Ave.
Chicago, IL 60637

Business Marketing
Crain Communications
1400 Woodbridge
Detroit, MI 48207

Byte
McGraw-Hill Inc.
One Phoenix Mill Lane
Peterborough, NH 03458

Cable Television Business
Cardiff Publishing
6300 S. Syracuse Way, Suite 650
Englewood, CO 80111

California Engineer
California Engineer Publ. Co.
221 Bechtel Engr. Ctr.
Berkeley, CA 94720

Car Craft
8490 Sunset Blvd.
Los Angeles, CA 90069

Casting World
Continental Communications
P.O. Box 1919
Bridgeport, CT 06601–1919

Certified Engineering Technician
American Society of Certified
Engineering Technicians
P.O. Box 371474
El Paso, TX 79937

*Chemical and Petroleum
Engineering*
Plenum Publishing Corp.
233 Spring St.
New York, NY 10013

Chemical Business
Schnell Publishing Co.
80 Broad St.
New York, NY 10004–2203

Chemical Engineering
McGraw-Hill
1221 Avenue of the Americas
New York, NY 10020

Chemical Engineering Progress
American Institute of Chemical
Engineers
345 E. 47th St.
New York, NY 10017

Chemical Processing
Penton Publishers
301 E. Erie St.
Chicago, IL 60611

Chief Engineer
Chief Engineers Association of Chicago
11340 W. 159th St.
Overland Park, IL 60462

Civil Engineering
American Society of Civil Engineers
345 East 47th St.
New York, NY 10017–2398

Coal
11 W. 19th St.
New York, NY 10011

Coal
Maclean Hunter Publishing Company
29 N. Wacker Dr.
Chicago, IL 60606

Combustion and Flame
Elsevier Science Publishing Co.
655 Avenue of the Americas
New York, NY 10010

Commodore Magazine
1200 Wilson Dr.
West Chester, PA 19380

Communication News
HBJ Publications, Inc.
124 S. First St.
Geneva, IL 60134

Compressed Air
253 E. Washington Ave.
Washington, NJ 07882–2495

Computer-Aided Engineering
Penton Pub.
1100 Superior Ave.
Cleveland, OH 44114

Computerworld
IDG Communications
375 Cochituate Rd. Box 9171
Framingham, MA 01701

Computer Design
Penwell
One Technology Park
Westford, MA 01886

Computer Graphics Today
Media Horizons Inc.
50 W. 23rd St.
New York, NY 10010

*Computers and Industrial
Engineering*
Pergamon Journals, Inc.
Maxwell House Fairview Park
Elmsford, NY 10523

Computers in Physics
American Institute of Physics
335 E. 45th St.
New York, NY 10017

Concrete Construction Magazine
426 S. Westgate
Addison, IL 60101–9929

Concrete Products
29 N. Wacker Dr.
Chicago, IL 60606

Constructor
1957 E. Street N.W.
Washington, DC 20006

Consulting Engineer
1350 E. Touhy, No. 5080
Des Plaines, IL 60018

Control Engineering
Cahners Publishing Company
1350 E. Touhy, No. 5080
Des Plaines, IL 60018

Cost Engineering
American Assn. of Cost Engineers
308 Monogahela Bldg.
Morgantown, WV 26505–5468

DE/Domestic Engineering
Delta Communications
385 N. York Rd.
Elmhurst, IL 60126

Defense Electronics
1170 E. Meadow Dr.
Palo Alto, CA 94303

Defense Science & Electronics
Rush Franklin Building
300 Orchard City Dr. Suite 234
Campbell, CA 95008

Designfax
International Thomson Indus. Press,
Inc.
6521 Davis Industrial Parkway
Solon, OH 44139

Design News
Cahners Publishing Co.
275 Washington St.
Newton, MA 02158–1630

Diesel & Gas Turbine Worldwide
13555 Bishop Court
Brookfield, WI 53005–6286

Dr. Dobbs Journal
M & T Publishing
501 Galveston Dr.
Redwood City, CA 94063

Drilling and Engineering
Society of Petroleum Engineers
P.O. Box 833836
Richardson, TX 75083–3836

EDN
Cahners Publishing Company
275 Washington St.
Newton, MA 02158–1630

EMC Technology
Interference Control Technologies,
Inc.
Rte 625, P.O. Box D
Gainesville, VA 22065

ENR Engineering News Record
McGraw-Hill Information Systems
1221 Avenue of the Americas
New York, NY 10020

Electric Light and Power
Pennwell Publishing Co.
1421 S. Sheridan Rd.
Tulsa, OK 74112

*Electrical Construction and
Maintenance*
McGraw-Hill Information Services
1221 Avenue of the Americas
New York, NY 10020

Electrical Wholesaling
1221 Avenue of the Americas
New York, NY 10020

Electrical World
McGraw-Hill
11 W. 19th St.
New York, NY 10011

Electronic Business
Cahners Publishing Co.
275 Washington St.
Newton, MA 02158-1630

Electronic Design
Hayden Publishing/VNU
10 Mulholland Dr.
Hasbrouck Hts., NJ 07604

Electronic/Electrical Product News
707 Westchester Ave.
White Plains, NY 10604

Electronic Engineering Times
CMP Publications
600 Community Dr.
Manhasset, NY 11030

Electronic Manufacturing
Lake Publishing Corp.
17730 W. Peterson Rd. Box 159
Libertyville, IL 60048

Electronic Packaging & Production
1350 E. Touhy Ave.
P.O. Box 5080
Des Plaines, IL 60017-5080

Electronic Products
645 Stewart Ave.
Garden City, NY 11530

Electronic Servicing & Technology
P.O. Box 12901
Overland Park, KS 66212-9981

Electronics
VNU Business Publications, Inc.
Ten Mulholland Dr.
Hasbrouck Heights, NJ 07604

Energy Management Technology
Walter-Davis Publications
2500 Office Center
Willow Grove, PA 19090

Engineered Systems
Business News Publishing Co.
P.O. Box 2600
Troy, MI 48007

Engineering Education
American Society for Engineering
Education
Eleven Dupont Circle Suite 200
Washington, D.C. 20036

Engineering and Mining Journal
300 W. Adams St.
Chicago, IL 60606

Engineering Tools
VNU Business Publications
10 Mulholland Dr.
Hasbrouck Heights, NJ 07604

Environmental Science & Technology
American Chemical Society
1155 16th St. N.W.
Washington, D.C. 20036

Equal Opportunity Publications
44 Broadway
Greenlawn, NY 11740

Evaluation Engineering
Nelson Publishing
2504 Tamiami Trail
Nokomis, FL 34275

Executive Computing
Association of Computer Users
P.O. Box 2189
Berkeley, CA 94702

Food Engineering
Chilton Company
Chilton Way
Radnor, PA 19089

Food Processing
301 E. Erie St.
Chicago, IL 60611

*Foundry Management and
Technology*
Penton Publishing Co.
1100 Superior Ave.
Cleveland, OH 44114

Geology
Geological Society of America
3300 Penrose
Boulder, CO 80301

Geotimes
American Geological Inst.
4220 King St.
Alexandria, VA 22302–1507

Graduating Engineer
McGraw-Hill
1221 Avenue of the Americas Ste 4360
New York, NY 10020

Ham Radio
Communications Technology, Inc.
Main St.
Greenville, NH 03048

Harvard Business Review
Soldiers Field
Boston, MA 02163

Heating/Piping/Air Conditioning
Penton Publishing Co.
1100 Superior Ave.
Cleveland, OH 44114

*Heating, Air Conditioning
& Plumbing Products*
P.O. Box 1952
Dover, NJ 07801–0952

High Performance Systems
600 Community Dr.
Manhasset, NY 11030

High Technology Careers
Westech Publishing Co.
4701 Patrick Henry Dr.
Santa Clara, CA 95054

Home Office Computing
730 Broadway
New York, NY 10003

Hybrid Circuit Technology
Lake Publishing Co.
17730 W. Peterson Rd. Box 159
Libertyville, IL 60048

Hydraulics and Pneumatics
Penton Publishing
1100 Superior Ave.
Cleveland, OH 44114

Hydrocarbon Processing
P.O. Box 2608
Houston, TX 77252

IAN-Instrument & Control News
Chilton Company
Chilton Way
Radnor, PA 19089

*I&CS-Control Tech. for Engrs. and
Engr. Mgmt*
Chilton Company
Chilton Way
Radnor, PA 19089

IEEE Spectrum
Inst. of Elec. and Electronic Engrs.
345 E 45th St.
New York, NY 10017

INTECH
Instrument Society of America
P.O. Box 12277
Research Triangle Park, NC 27709

Industrial Chemist
McGraw-Hill
1221 Avenue of the Americas
New York, NY 10020

Industrial Design
Design Publications, Inc.
330 W. 42nd St.
New York, NY 10036

Industrial Distribution
Cahners Publishing Co.
275 Washington St.
Newton, MA 02158

Industrial Engineering
Institute of Industrial Engineers
25 Technology Park/Atlanta P.O. Box
6510
Norcross, GA 30091–6150

Industrial Finishing
Hitchcock Building
Wheaton, IL 60188

Industrial Heating
National Industrial Publishing Co.
1000 Killarney Dr.
Pittsburgh, PA 15234

*Industrial Maintenance &
Plant Operation*
Chilton Company
Chilton Way
Radnor, PA 19089

Industrial Photography
210 Crossways Park Dr.
Woodbury, NY 11797

Industrial Safety & Hygiene News
Chilton Company
Chilton Way
Radnor, PA 19089

Information Display
201 Varick St. Suite 1140
New York, NY 10014

Infoworld
1060 Marsh Rd.
Menlo Park, CA 94025

Iron and Steel Engineer
Three Gateway Center Suite 2350
Pittsburgh, PA 15222

JAPCA
Three Gateway Center, Four West
Pittsburgh, PA 15230

Journal of Accountancy
1211 Avenue of the Americas
New York, NY 10036

Journal of Electronic Defense
Horizon House
685 Canton St.
Norwood, MA 02062

Journal of Petroleum Technology
Society of Petroleum Engineers
P.O. Box 833836
Richardson, TX 75083–3836

*Journal of Small Business
Management*
West Virginia University
P.O. Box 6025
Morgantown, WV 26506–6025

Journal of Systems Management
24587 Bagley Rd.
Cleveland, OH 44138

Laboratory Management
Nature Publishing Co.
65 Bleecker St.
New York, NY 10012–2420

Laser Focus
Penn Well Publishing
One Technology Park Drive
Westford, MA 01886

Lasers and Optronics
High Tech Pubs
P.O. Box 650
Morris Plains, NJ 07950

Lubrication Engineering
Society of Tribologists & Lubrication
Engineers
838 Busse Highway
Park Ridge, IL 60068

Manage
2210 Arbor Blvd.
Dayton, OH 45439

Managing Automation
Thomas Publishing
One Penn Plaza
New York, NY 10001

Manufacturing Engineering
One SME Dr.
P.O. Box 930
Dearborn, MI 48121

Material Handling Engineering
Penton Publishing
1100 Superior Ave.
Cleveland, OH 44114

Materials Engineering
Penton Publishing
1100 Superior Ave.
Cleveland, OH 44114

Mechanical Engineering
American Society of Mechanical
Engineers
345 E. 47th St.
New York, NY 10017

*Medical Electronics and Equipment
News*
532 Busse Highway
Park Ridge, IL 60068

Metalworking News
Fairchild Publications
Seven E. 12th St.
New York, NY 10003

Metalworking Digest
Gordon Publications Inc.
Box 1952
Dover, NJ 07801

MicroCAD News
Ariel Communications
12710 Research, Suite 250
P.O. Box 203550
Austin, TX 78759

Microtimes
BAM Publications
5951 Canning St.
Oakland, CA 94609

Microwave Journal
685 Canton St.
Norwood, MA 02062

Microwave News
P.O. Box 1799
Grand Central Station
New York, NY 10163–1799

Microwaves and RF
VNU Business Publications
10 Mulholland Dr.
Hasbrouck Heights, NJ 07604

The Military Engineer
Society of American Military Engineers
607 Prince St.
P.O. Box 21289
Alexandria, VA 22320–2289

Mini-Micro Systems
Cahners Publishing Co.
275 Washington St.
Newton, MA 02158–1630

Mining Engineering
Society of Mining Engineers, Inc.
P.O. Box 625002
Littleton, CO 80162–5002

Modern Casting
American Foundrymen's Society
Golf & Wolf Rds.
Des Plaines, IL 60016–2277

Modern Electronics
CQ Commun., Inc.
76 N. Broadway
Hicksville, NY 11801

Modern Machine Shop
6600 Clough Pike
Cincinnati, OH 45244

Modern Materials Handling
Cahners Publishing
275 Washington St.
Newton, MA 02158

Modern Metals
Delta Communications, Inc.
400 N. Michigan Ave.
Chicago, IL 60611

Modern Plastics
1221 Avenue of the Americas
New York, NY 10020

Modern Railroads
Int. Thomson Transport Press
424 33rd St.
New York, NY 10001

National Engineer
National Association of Power
Engineers
2350 E. Devon Ave. Suite 115
Des Plaines, IL 60018

National Petroleum News
950 Lee St.
Des Plaines, IL 60016

National Public Accountant
1010 N. Fairfax St.
Alexandria, VA 22314

National Safety and Health News
National Safety Council
444 N. Michigan Ave.
Chicago, IL 60611

Nibble
52 Domino Dr.
Concord, MA 01742

Nuclear News
American Nuclear Society
555 N. Kensington Ave.
La Grange Park, IL 60525

Nuclear Times
1601 Conn. Ave. N.W. No. 300
Washington, DC 20009

Oil and Gas Journal
Penwell Publishing
P.O. Box 1260
Tulsa, OK 74101

Packaging
Cahners Publishing Company
1350 E. Touhy No. 5080
Des Plaines, IL 60017–5080

Paper, Film & Foil Converter
MacLean Hunter Publishing
29 N. Wacker Dr.
Chicago, IL 60606

Paper Trade Journal
Vance Publishing
400 Knightsbridge Pkwy.
Lincolnshire, IL 60069

Personal Computing
VNU Business Pub., Inc.
10 Mullholland Dr.
Hasbrouck Hts., NJ 07604

Petroleum Engineer International
Edgell Communications Inc.
7500 Old Oak Rd.
Cleveland, OH 44130

Photomethods
1090 Executive Way
Des Plaines, IL 60018

Pipeline & Gas Journal
Edgell Communications Inc.
7500 Old Oak Rd.
Cleveland, OH 44130

Pipe Line Industry
3301 Allen Parkway Box 2608
Houston, TX 77001

Plant Engineering
Cahners Publishing Company
249 W. 17th St.
New York, NY 10011

Plant Services
301 E. Erie St.
Chicago, IL 60611

Plastics Engineering
Society of Plastic Engineers
14 Fairfield Dr.
Brookfield, CT 06804

Plastics Technology
110 N. Miller Rd.
Akron, OH 44313

Plastics World
Cahners Publishing Company
275 Washington St.
Newton, MA 02158

Power
McGraw-Hill
11 W. 19th St.
New York, NY 10011

Power Engineering
1421 Sheridan Rd.
Tulsa, OK 74112

Power Transmission Design
Penton Publishers
1100 Superior Ave.
Cleveland, OH 44114

Precision Metal
Penton Publishing Co.
1100 Superior Ave.
Cleveland, OH 44114

Private Pilot
Fancy Publications, Inc.
Box 6050
Mission Viejo, CA 92690

Product Design and Development
Chilton Co.
Chilton Way
Radnor, PA 19089

Programmable Controls
ISA Services, Inc.
67 Alexander Dr.
Box 12277
Research Triangle Park, NC 27707

Public Power
2301 M St., N.W.
Washington, D.C. 20037

Pulp and Paper
500 Howard St.
San Francisco, CA 94105

QST, ARRL
225 Main St.
Newington, CT 06111

Quality
Hitchcock Publishing
25W550 Geneva Rd.
Wheaton, IL 60187

RF Design
Cardiff Publishing Co.
6300 S. Syracuse Way, Suite 650
Englewood, CO 80111

Radio-Electronics
500-B Bi-County Blvd.
Farmingdale, NY 11735

Radiology Today
Slack, Inc.
6900 Grove Rd.
Thorofare, NJ 08086-9447

Railway Age
345 Hudson St.
New York, NY 10014

Research and Development
Cahners Publishing Company
249 W. 17th St.
New York, NY 10011

Rock Products
29 N. Wacker St.
Chicago, IL 60606

Rural Electrification
1800 Massachusetts Ave. N.W.
Washington, D.C. 20036

SAMPE Journal
Society for the Advancement of
Material
P.O. Box 2459
Covina, CA 91722

SMPTE Journal
Society of Motion Picture and
Television Engineers
595 W. Hartsdale Ave.
White Plains, NY 10607

Safety & Health
National Safety Council
444 N. Michigan Ave.
Chicago, IL 60611

Science
American Assoc. for the
Advancement of Science
1333 H St. N.W.
Washington, D.C. 20005

Science & Technology
48 E. 43rd St.
New York, NY 10017

Science Education
John Wiley & Sons, Inc.
605 Third Ave.
New York, NY 10158

Science News
1719 N Street N.W.
Washington, DC 20036

Scientific American
415 Madison Ave.
New York, NY 10017

Sea Technology
Compass Publications
1117 N. 19th St., Suite 1000
Arlington, VA 22209

Sky and Telescope
P.O. Box 9111
Belmont, MA 02178-9111

Solid State Technology
14 Vanderventer Ave.
Port Washington, NY 11050

*Space Technology: Industrial &
Commercial Applications*
Pergamon Journals, Inc.
Maxwell House Fairview Park
Elmsford, NY 10523

Spectroscopy
Aster Publishing Co.
859 Willamette St.
P.O. Box 10955
Eugene, OR 97440

Supervision
424 N. 3rd St.
Burlington, IA 52601-5224

Telecommunications
685 Canton St.
Norwood, MA 02062

Test & Measurement World
Cahners Publishing Co.
275 Washington St.
Newton, MA 02158

Test Engineering & Management
The Mattingley Pub. Co., Inc.
3756 Grand Ave., Suite 205
Oakland, CA 94610

The Physics Teacher
American Association of Physics
Teachers
5112 Berwyn Rd. 2nd Floor
College park, MD 20740

Today's Chemist
American Chemical Society
500 Post Road E.
P.O. Box 231
Westport, CT 06881

Tooling and Production
6521 Davis Industrial Parkway
Solon, OH 44139

Traffic Safety
National Safety Council
444 N. Michigan Ave.
Chicago, IL 60611

Traffic World
1325 G St. N.W. Suite 900
Washington, DC 20005

Transmission & Distribution
Andrews Communications, Inc.
5123 W. Chester Pike
P.O. Box 556
Edgemont, PA 19028

TV Technology
5827 Columbia Pike, Suite 310
Falls Church, VA 22041

Water Engineering and Management
Scranton Gilette Communications, Inc.
380 Northwest Highway
Des Plaines, IL 60016

Welding Journal
American Welding Society
Box 351040
Miami, FL 33125

World Oil
Gulf Publishing Do.
Box 2608
Houston, TX 77252

Appendix C
Addresses of
On-Line Vendors

American Open University
New York Institute of Technology
Central Islip, NY 11722
1-800-222-6948

BRS Information Technologies
1200 Route 7
Latham, NY 12110
1-800-468-0908

Chemical Abstracts Service
A Division of the American Chemical
Society
2540 Olentangy River Rd.
P.O. Box 3012
Columbus, OH 43210-0012
614-447-3600
1-800-848-6538: Customer Service
1-800-848-6533: STN Search Assistance Desk

CompuServe Information Services
5000 Arlington Centre Blvd.
P.O. Box 20212
Columbus, OH 43220
614-457-8650
1-800-848-8990

Connect.ed
92 Van Cortlandt Park South, #6F
Bronx, NY 10463
212-549-6509

Delphi
3 Blackstone St.
Cambridge, MA 02139
1-800-544-4005
617-491-3393

Dialog Information Services, Inc.
3460 Hillview Ave.
Palo Alto, CA 94304
415-858-3785
1-800-3-DIALOG

Dow Jones
P.O. Box 300
Princeton, NJ 08543-0300
609-452-1511

GEnie
General Electric Information Services
401 N. Washington St.
Rockville, MD 20850
301-340-4000
1-800-638-9636

Knowledge Index
415-858-3785
1-800-334-2564

Mead Data Central
9443 Springboro Pike DM
P.O. Box 933
Dayton, OH 45401
513-865-6800
1-800-227-4908

NEWSNET
945 Haverford Rd.
Bryn Mawr, PA 19010
800-537-0808 in PA
1-800-345-1301

Nova University
3301 College Ave.
Fort Lauderdale, FL 33314
1-305-475-7047

Orbit Search Service
8000 Westpark Dr.
McLean, VA 22102
703-442-0900
1-800-456-7248

Prodigy
445 Hamilton Ave.
White Plains, NY 10601
914-993-8000

Videolog Communications
50 Washington St.
Norwalk, CN 06854
203-838-5100
1-800-843-3656

VU/TEXT Information Services, Inc.
325 Chestnut St. Suite 1300
Philadelphia, PA 19106
215-574-4416
1-800-323-2940

Appendix D
Exchange Sort

A variation to the method of chapter 5, known as an *exchange sort,* is covered here. To demonstrate this method, the identical cards and numbers used in chapter 5 will be used.

Place the cards in a single row as illustrated in FIG. D-1. To begin the procedure, compare the first card, A-1941, with the second card, B-187. B-187 is smaller than A-1941, so exchange their positions, as illustrated in FIG. D-2.

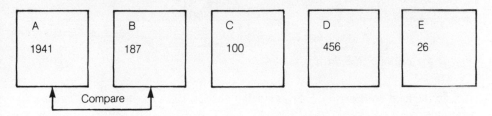

Fig. D-1. Original order.

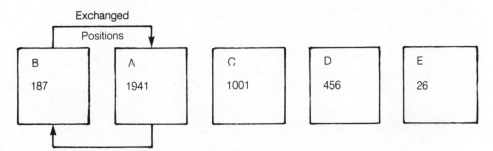

Fig. D-2. Results of the first comparison.

For the second step, compare A-1941 with C-1001. Since C-1001 is smaller, exchange its position with A-1941, as illustrated in FIG. D-3.

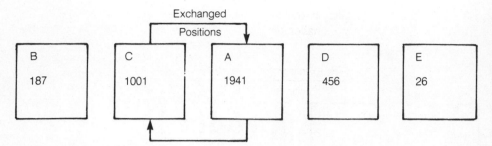

Fig. D-3. Results of the second comparison.

For the next comparison, note that D-456 is smaller than A-1941, so exchange their positions as shown in FIG. D-4.

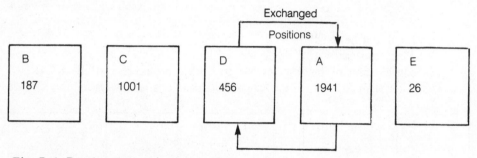

Fig. D-4. Results of the third comparison.

In the next comparison, E-26 is smaller than A-1941 so the two exchange positions as shown in FIG. D-5.

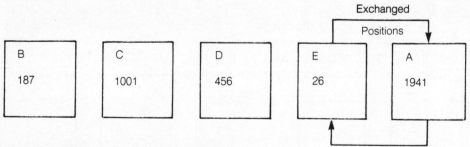

Fig. D-5. Results of the fourth comparison.

So, at the end of the first round of comparisons, the order is as shown in FIG. D-5. Note that the largest number, A-1941, has moved to the rightmost position. Also note that four ($n-1$) comparisons were needed to sort five (n) cards.

To begin the second round of comparisons, compare B-187 with C-1001.

Since B-187 is smaller, it remains in its position. Next compare C-1001 with D-456. D-456, being smaller, exchanges positions with C-1001 as shown in FIG. D-6.

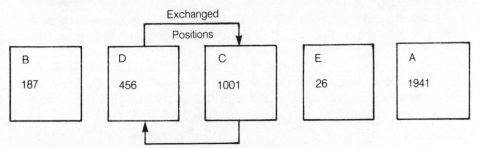

Fig. D-6. Results of the second comparison, Round 2.

For the next comparison, E-26 and C-1001 exchange positions, as shown in FIG. D-7. Note that since the largest number, A-1941, has already moved to the rightmost position, a comparison with C-1001 is not necessary.

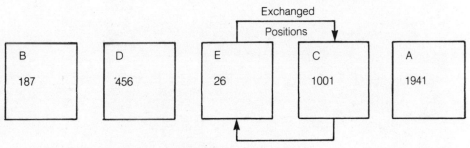

Fig. D-7. Results of the third comparison, Round 2.

The remainder of the comparisons are detailed here: Compare B-187 with D-456. No exchange. Compare D-456 with E-26. Exchange their positions. Compare B-187 with E-26. Exchange their positions. The final order, after four complete rounds of comparisons, is shown in FIG. D-8.

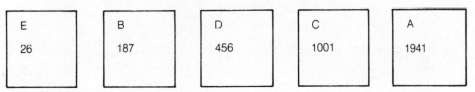

Fig. D-8. The final results—Sorted order.

Practice both this method and the method described in chapter 5 to see which one works the best for you. One disadvantage of the open-card method discussed here is that all the cards are face up and might distract you when you try to make comparisons. When you use this method, it might work best if you pick up the two cards you are comparing so you can concentrate on the two topics without being distracted.

Another possible variation, particularly if you have a large number of topic cards, is to place the cards in one stack and go through the deck, top to bottom repeatedly. Compare one pair of the topics at a time and exchange their positions if one topic is to be covered before the other. Keep exchanging their order as you work through the deck until all of the cards are in the desired order. Whichever method you use, although the mechanics might differ, the results will be the same.

Glossary

abstract A brief summary of an article, book, or report, which also lists author, title, publisher, and date of publishing.

acoustic coupler An early type of modem with rubber cups that cradle over the handset of a telephone.

agate line A unit of measure used in selling newspaper advertising space. Fourteen agate lines equal one column inch.

ampersand The word used to denote "&," the symbol for the word *and.*

analog signals The signals carried over telephone lines.

ASCII code The standard code developed by the American Institute for Information Interchange that is used for all communications and represents letters, numbers, and symbols as bit patterns. ASCII files can be read by practically all software.

asynchronous Of or referring to a method of transmitting signals in which data is sent one bit at a time.

autodial A modem feature that lets the computer automatically dial a telephone number that is stored in the computer's memory.

back matter The printed matter following the end of the text of a book.

basis weight The weight of a ream (500 sheets) of paper. One pound of 20 lb. paper weighs 20 lbs.

baud rate A technical term referring to the rate at which data is transmitted over telephone lines. In most cases, the baud approximately equals bits per second.

bibliography A list of reference literature located in the back matter of a book or at the end of an article or chapter.

boldface Type that is heavier than the text type with which it is used.

BPS Abbreviation for bits per second, the speed at which data is transmitted.

buffer Computer memory set aside to store information obtained from an on-line service.

bullet A solid dot character used ornamentally.

camera-ready   Of or referring to a pasteup that can be photographed as is to produce a negative for plate-making, without the need to typeset text or redraw artwork.

contact print   A photographic print made with a negative or positive in contact with sensitized paper. No camera is necessary. Images are reversed, as from negative to positive, and prints can only be made the same size as the original.

cropping   Marking a drawing or photograph to indicate that only a portion of the artwork is to be used, instead of physically cutting out the desired area. Usually, the entire piece of art is actually reproduced as film and the undesired image area then cut away.

direct connect   A design that allows a modular telephone cord to connect directly into a phone line.

downloading   The process of receiving information from one computer and storing it on another.

electronic mail   A service that lets you send messages to other users on the same on-line system.

elite   A type font that has 12 characters per linear inch and 6 lines to the vertical inch.

errata   A list of errors, corrections, and deletions discovered after an article or book has been printed.

exploded view   An illustration that displays separate parts of a component displaced outward in order of disassembly.

flush   Even with margins. For example, for flush right all characters line up vertically at the right-hand margin.

font   A complete alphabet of any one typeface in a given point size — uppercase, lowercase, numerals, punctuation marks, etc.

front matter   All matter preceding the first page of text.

full duplex   Simultaneous transmission of data in both directions.

galley proofs   A preliminary reproduction of text composition for the purpose of checking spelling, spacing, etc. before pasteup or makeup.

half duplex   The transmission of data in only one direction at a time.

halftone   A photo, engraved plate, or printed illustration in which a range of solid tones is reproduced by a pattern of dots of varying sizes.

hardcopy   A printed copy of writing.

Hayes-compatible   Of or referring to modems that use commands originated by Hayes Microcomputer Products. Hayes compatibility is not an absolute requirement, but is a de facto industry standard.

head, heading A title or caption at the head of a chapter, section, column, list, table, or illustration.

line drawing A drawing made from lines with no gradation of gray tones and no dot pattern.

log-on The process of accessing an on-line service.

masthead The matter printed in each issue of a periodical stating the ownership, title, editorial staff, etc.

modem A contraction of *mo*dulator/*dem*odulator, the interface box or card that converts a signal into two different frequencies so that it can be transmitted over telephone lines and that also performs the reverse function, converting the signal on the phone lines to ones that can be handled by a computer. The modem converts the computer's 1's and 0's into two musical tones that whistle down the telephone line and are turned back into 1's and 0's at the receiving end.

parity A setting for an error-checking bit during transmission of data.

pica A standard type font that has ten characters per inch.

point A printer's measure equal to $\frac{1}{72}$ inch.

protocol The agreed upon settings for the transfer of information between systems. Typical protocols are XMODEM, Y-MODEM, Z-MODEM, CompuServe-B, Kermit, and ASCII. The two computers must agree on parity and protocol for them to be able to interchange information.

recto A right-hand, odd-numbered page.

RS-232C The standard established by the Electronic Industries Association for serial transmission of data for telecommunications.

running head A title repeated at the top of each page of a book.

soft copy A copy of written material that is stored in digital form on a diskette or other storage media. A printer converts soft copy to hardcopy.

stet A printer's term meaning "Let it stand," used in editing to retain material that has been crossed off.

stub The list of subjects or entries at the left of a table.

subhead A secondary headline or title.

subtitle A secondary title, often an explanation or expansion of the main title.

synchronous Of or referring to a system of transmission in which data is sent in blocks and the receiver and sender must be in synchronism.

terminal A device that can send and receive information from a computer.

Glossary

uploading   Transmitting information locally and storing it at a remote location.

verso   A left-hand, even-numbered page.

widow   A single word or partial line of words spilling over from the previous page and appearing at the top of the next page.

Bibliography

Adams, James L. *The Care & Feeding of Ideas.* Reading, MA: Addison-Wesley, 1986.

Alley, Michael. *The Craft of Scientific Writing.* Englewood Cliffs, NJ: Prentice Hall, 1987.

Blicq, R.S. *Technically — Write!* Englewood Cliffs, NJ: Prentice Hall, 1972.

Dodds, Robert H. *Writing for Technical and Business Magazines.* New York: John Wiley & Sons, 1969.

Ehrlich, Eugene, and Daniel Murphy. *The Art of Technical Writing.* New York: Bantam Books, 1964.

Emerson, Connie. *Write on:Target,* Cincinnati, OH: Writer's Digest, 1981.

Ferrarini, Elizabeth. *Infomania.* Boston, MA: Houghton Mifflin, 1985.

Glossbrenner, Alfred. *How to Look it up Online.* New York: St. Martins Press, 1987.

——————. *Personal Computer Communications.* New York: St. Martins Press, 1985.

Glover, John A. *Becoming a More Creative Person.* Englewood Cliffs, NJ: Prentice-Hall, 1980.

Hicks, Tyler G. *Writing for Engineering and Science.* New York: McGraw-Hill, 1961.

Helliwell, John. *Inside Information,* New York: New American Library, 1986.

Bibliography

Hoover, Hardy. *Essentials for the Scientific and Technical Writer.* New York: Dover Publications, 1980.

Li, Tze-chung. *An Introduction to Online Searching.* Westport, CT: Greenwood Press, 1985.

Mills, Gordon H. and John A. Walter. *Technical Writing.* New York: Holt, Rhinehart and Winston, 1962.

Mullins, Carolyn J. *The Complete Writing Guide,* Englewood Cliffs, NJ: Prentice-Hall, 1980.

Sherman, Theodore A. and Simon S. Johnson. *Modern Technical Writing.* New York: Prentice Hall, 1975.

Tichy, H.J. *Effective Writing.* New York: John Wiley & Sons, 1966.

Turner, Barry T. *Effective Technical Writing and Speaking.* London: Business Books, 1978.

Index

Index

Other Bestsellers of Related Interest

ENCYCLOPEDIA OF ELECTRONICS
—2nd Edition—Stan Gibilisco and
Neil Sclater, Co-Editors-in-Chief
Praise for the first edition:
*". . . a fine one-volume source of detailed
informationfor the whole breadth of electronics."*

—Modern Electronics
The second edition, newly revised and
expanded, brings you more than 950 pages of
listings that cover virtually every electronics
concept and component imaginable. From
basic electronics terms to state-of-the-art
applications, this is the most complete and
comprehensive reference available for anyone involved in any area of electronics practice! 976 pages, 1400 illustrations, Book No.
3389, $68.95 hardcover only.

AutoCAD PROGRAMMING—
Dennis N. Jump
CAD expert Dennis Jump offers you a
straightforward, comprehensive guide to the
popular CAD software package that includes
version 2.0. Jump explains in detail the data
structures and algorithms associated with
AutoCAD programming, including numerous sample program listings in both C and
BASIC. You'll learn how to write application
programs that use AutoCAD as a companion, as well as how to use data within Auto-
CAD to display the images, drawings,
diagrams, and more created with your application programs. 288 pages, 150 illustrations, Book No. 3093, $24.95 paper, $33.95
hardcover.

**UNDERSTANDING DIGITAL
ELECTRONICS**
—2nd Edition—R.H. Warring and Michael
J. Sanfilippo
This revised edition of the bestselling
guidebook to digital electronics is the perfect
tool to help you keep up with the growth and
change in technology. It's a quick and complete resource of all the principles and concepts of digital circuits, providing coverage
of important areas such as binary numbers,
digital logic gates, Boolean algebraic theo-
rems, flip-flops and memories, number systems, and arithmetic logic units (including
the 74181 ALU). 196 pages, 172 illustrations, Book No. 3226, $14.95 paperback,
$22.95 hardcover.

**COMPUTER TECHNICIAN'S
HANDBOOK**
—3rd Edition—Art Margolis
*"This is a clear book, with concise and sen-
sible language and lots of large diagrams . .
. use [it] to cure or prevent problems in
[your] own system . . . the [section on trou-
bleshooting and repair] is worth the price of
the book."* *—Science Software Quarterly*
MORE than just a how-to manual of do-
it-yourself fix-it techniques, this book offers
complete instructions on interfacing and
modification that will help you get the most
out of your PC. 580 pages, 97 illustrations,
Book No. 3279, $24.95 paperback, $36.95
hardcover only.

Other Bestsellers of Related Interest

ELECTRONIC DATABOOK—
Fourth Edition—Rudolf F. Graf

If it's electronic, it's here—current, detailed, and comprehensive! Use this book to broaden your electronics information base. Revised and expanded to include all up-to-date information, the fourth edition of *Electronic Databook* will make any electronic job easier and less time-consuming. This edition includes information that will aid in the design of local area networks, computer interfacing structure, and more! 528 pages, 131 illustrations, Book No. 2958, $24.95 paperback, $34.95 hardcover.

ELECTRONICS EQUATIONS HAND BOOK—Stephen J. Erst

Here is immediate access to equations for nearly every imaginable application! In this book, Stephen Erst provides an extensive compilation of formulas from his 40 years' experience in electronics. He covers 21 major categories and more than 600 subtopics in offering the over 800 equations. This broadbased volume includes equations in everything from basic voltage to microwave system designs. 280 pages, 219 illustrations, Book No. 3241, $16.95 paperback, $24.95 hardcover.

INVENTING: Creating and Selling Your Ideas—Philip B. Knapp, Ph.D.

If you've ever said, "Somebody ought to invent a (so-and-so) that will do (such-and-such) . . ." you should read this guide. Philip B. Knapp supplies valuable advice and practical guidance on transforming your ideas into working items and selling them! 250 pages, Book No. 3184, $15.95 paperback, $24.95 hardcover.

THE COMPLETE HANDBOOK OF MAGNETIC RECORDING—
3rd Edition—Finn Jorgensen

This book covers virtually every aspect of the magnetic recording science. A recognized classic in its field, this comprehensive reference makes extensive use of illustrations, line drawings, and photographs. Audio recording, instrumentation recording, video recording, FM and PCM recording, as well as the latest digital techniques are thoroughly described. 768 pages, 565 illustrations, Book No. 3029, $44.50 hardcover only.

UNDERSTANDING LASERS—
Stan Gibilisco

If you could have only one book that would tell you everything you need to know about lasers and their applications—this would be the book for you! Covering all types of laser applications—from fiberoptics to supermarket checkout registers—Stan Gibilisco offers a comprehensive overview of this fascinating phenomenon of light. He describes what lasers are and how they work, and examines in detail the different kinds of lasers in use today. 180 pages, 96 illustrations, Book No. 3175, $14.95 paperback, $23.95 hardcover.

Look for These and Other TAB Books at Your Local Bookstore

To Order Call Toll Free 1-800-822-8158

(in PA, AK, and Canada call 717-794-2191)

or write to TAB BOOKS, Blue Ridge Summit, PA 17294-0840.

Title	Product No.	Quantity	Price

☐ Check or money order made payable to TAB BOOKS

Charge my ☐ VISA ☐ MasterCard ☐ American Express

Acct. No. _____ Exp. _____

Signature: _____

Name: _____

Address: _____

City: _____

State: _____ Zip: _____

Subtotal $ _____

Postage and Handling
($3.00 in U.S., $5.00 outside U.S.) $ _____

Add applicable state
and local sales tax $ _____

TOTAL $ _____

TAB BOOKS catalog free with purchase; otherwise send $1.00 in check or money order and receive $1.00 credit on your next purchase.

Orders outside U.S. must pay with international money order in U.S. dollars.

TAB Guarantee: If for any reason you are not satisfied with the book(s) you order, simply return it (them) within 15 days and receive a full refund. **BC**